Recipes to Reconnect

Anna Boglione

An Hachette UK Company
hachette.co.uk

First published in Great Britain in 2023
by Kyle Books, an imprint of
Octopus Publishing Group Limited

Carmelite House
50 Victoria Embankment
London EC4Y 0DZ
kylebooks.com

ISBN: 978 0 85783 9961

Distributed in the US by Hachette Book Group,
1290 Avenue of the Americas,
4th and 5th Floors, New York, NY 10104

Distributed in Canada by Canadian Manda Group,
664 Annette St., Toronto, Ontario,
Canada M6S 2C8

Publisher: Joanna Copestick
Editor: Isabel Jessop
Copy-editor: Katherine Delves
Art direction and design concept: Carol Montpart Studio
Graphic design: Andrin Rohner
Creative direction and Photography: Anna Boglione
Production: Katherine Hockley

A cataloguing in publication record for this
title is available from the British Library.

Printed and bound in China
10 9 8 7 6 5 4 3 2 1

RECIPES TO RECONNECT

Food and Conversations to Re-establish
our Relationship with Nature, Food and Self

Anna Boglione

Kyle Books

AUTUMN

WINTER

For my
niece and nephews,
children
to come and those
on their way.

When interviewing my brother about regenerative farming, my nine-year-old nephew came in with a note, asking if he could say something for the book.

RAFFI
I was going to say, if we keep fishing like we are now, in forty years there'll be no more fish in the ocean. There'll be more plastic than fish in the ocean.

HARRY
It's a scary fact.

ANNA
It is a scary fact. You know what I thought was very interesting is that during this time with COVID everybody stopped. It showed that it doesn't take much to regenerate ecosystems. The fact that everybody did stop for those months and those periods shows that we can have a really big impact on the environment if we all put our heads together and do it as one.

HARRY
Hopefully, your generation will make an impact. We're working hard on it, but there are lots of idiots in our generation.

INTRODUCTION

Over the last fifteen years, I have invested much of my time into learning and exploring the world of gut health and the connection it has with both mental and physical well-being and the natural world. My digestive issues started as a teenager when my vegetarian self was let loose in the world. As most uneducated teens do, I ate badly. Having grown up eating organic, eating balanced meals, I was now filling my tummy with heavily processed gluten – supermarket sandwiches were my go-to. As my gut health deteriorated, inflammation took hold, my stomach swelled and at times it became too painful to eat. It was my first inkling that you are what you eat. Gluten had made (and still does make) tiny holes in my intestinal tract, causing leaky gut. Often leaky gut can affect your brain function and, in my case, it gave way to brain fog in a serious sense. Brain fog can be crippling, affecting short-term memory, cognitive function and the ability to put words together. Labelled as dyslexic, dyspraxic and having attention deficit disorder, as a child I was stamped with adjectives for conditions I didn't fully understand. The added brain fog made me further question my intelligence. Phonetic decoding in dyslexics is on average five times as long as others – my first major hurdle in life was reading 'The fat cat sat on the mat'. Dyslexics can feel a high level of shame around their intelligence and reading age, often finding it easier to shy away from school. On average, thirty percent of dyslexics drop out of school; for good measure, I decided to drop out twice. Truancy, shame and anxiety go hand in hand with digestive issues, but it was only much later in life that I'd look back and see that emotional trauma had played a part in the deterioration of my physical health. At times hanging on to a slither of myself, I decided that university was not for me.

Reconnecting with my creativity, I set up an interactive production company, Petersham Road, merging promenade and immersive theatre. We specialized in site-specific entertainment, innovative music platforms and curated sensory experiences. My theatrical partner, Louis Waymouth, and I worked in my family's restaurant Petersham Nurseries during the day, and at night we set to planning our theatrical debut. Transforming Petersham House by filling it with performance and set, we would draw our guests from the mundane and into alternate realities. By creating a sense of intimacy within the narrative, the immersive aspect defined our style. We encouraged the spectator to integrate within the story. It is this all-encompassing, circular notion that still seeps into everything I do. Ten years flew by and I was burnt out. All that time I had been researching and trying to heal my gut (and mental health). I had tried and failed, and sometimes succeeded, with restrictive diets, remedies, potions, talking and movement therapies. I slowly found my way, with guidance and advice from wonderfully witchy practitioners, friends, relatives, books and talks. Having taken a pause from events, my thoughts came back to an interconnected solution and lifestyle.

Closing one events company, I opened another with food and holistic practices at its core. THE GUT is an events company that brings intricate topics to the table, in a digestible format, to start conversations, ignite curiosity and inspire change. We work with traditional and cutting-edge practitioners to heal on and off the mat. Guest chefs and I design meals with the highest quality local produce to complement a talk, supper club, retreat or workshop. For a long time, my mind has been tying threads between my health, diet, farming, food processes and nature. I spent my time researching and talking to experts on subjects to understand how we can achieve a holistic future. This research was the catalyst for starting THE GUT and also for filling these pages. The gut acts as a metaphor for connectivity, tangibly linking us to the world around us, to nature, to the soil and to ourselves – our personalities – our physical and emotional well-being. A spider diagram of interconnected wellness spread across my page, highlighting the overarching topics: nature, gut and mental health. Farming systems, land management and biodiversity not only have a key environmental role to play but also affect our own health. It is said that forty percent of the global population suffers from gut-related issues. Hippocrates, the father of modern medicine, famously exclaimed 'all disease begins in the gut'. Our microbiome, a lot of which is concentrated in our gut, has a whole array of roles, from regulating our immune and digestive systems to protecting against disease and even influencing our personality. Depleted soil will produce less nutritious plants and a depleted microbiome can have an effect on our mental health.

Nature needed to be a key component in writing and researching *Recipes to Reconnect*. Two years on from the global pandemic, change on the horizon seems a distant memory. Global temperature is on the rise and with that species are being wiped out and natural disasters are devastating communities. Monoculture landscapes have pushed wildlife into smaller pockets and dramatically altered the soil. Human activities, such as land management, deforestation, over-grazing and chemical farming, have been the cause of a large percentage of our planet's fertile land becoming desertified. No longer able to lock in carbon, some scientists believe this is cataclysmic for climate change. We see ourselves as separate from nature, but for five million years humans existed within natural habitats. It is only in the last several hundred that urban environments have evolved, gradually building walls between ourselves and the ecosystems that sustain us. In the conversations that follow, nearly all touch upon our connections to the natural world.

Activist, thinker and peace walker Satish Kumar and I discuss our connection to nature, how we are one and how nature needs to be part of our

GDP. I interview my farmer younger brother, Harry Boglione, on soil and regenerative farming practices, and how they can sequester carbon and encourage ecosystems. Dr Vandana Shiva, anti-GMO warrior, talked about how we need to reconnect with ancient knowledge, the importance of swapping seeds and myths around organic farming being land-hungry. Author and rewilder Isabella Tree and I discussed what we can do as individuals and communities to encourage wildlife into our backyards. With Merlin Sheldrake, author and mushrooms expert, I learned about the importance of fungi, their potentially revolutionary role in our planet's future and what they can teach us about connectivity.

Fungi, farming systems and soil can all have an impact on our health and microbiome and from here my spider diagram went straight back to the digestive system. The intestinal tract is where my initial interest in interconnected wellness began. The industrialization of our most precious commodity – food – saw a shift in environmental and human health. A prime example of this is wheat. From the way that the grain is milled to genetic modification, chemical farming and the additives that keep bread soft for weeks, wheat has been dramatically altered. However, as much as inflammation can be about diet, it is also about emotion and even how we hold our bodies. I spoke to author and nutritionist Hannah Richards about some unexpected causes of inflammation, ones that had never crossed my mind. The deeper I moved into adulthood, the more my symptoms shifted and morphed. At night, chronic pain gripped my spine and years went by without it subsiding. Fasting has been known to reduce inflammatory diseases, so I decided to start once a week for twenty-four to thirty-six hours. It changed my life. Buchinger is a leading clinic for therapeutic fasting, with over 100 years of experience. Its director, physician and fasting expert Dr Françoise Wilhelmi de Toledo, agreed to talk to me about our hunter-gather ancestors and how fasting is a natural process. The Ayurvedic philosophy embodies many themes that run through this book, using herbs, diet, lifestyle and environment to balance and heal the self. Ayurvedic practitioner, and author of many books on the holistic nature of this philosophy, Dr Robert Svoboda and I talked about keeping equilibrium within the body. A conversation with science writer and microbiome expert Dr Alanna Collen opened my mind to how deeply our gut microbes can control our physical and mental health. Often, unresolved ailments within the body come back to internalized anxieties and trauma – mental health extends far beyond our cognitive awakening. Adopting a belief in a higher self and reconnecting with nature not only allows different levels of positivity and gratitude but also helps us find a 'state of flow'. A state of flow is when you connect with your parasympathetic nervous system through non-hedonistic pleasure and activities. The person performing the activity is fully involved, energized and focused, feeling pleasure, satisfaction and enjoyment – 'Entrenched in my writings, I reach a state of flow'. Meditation deactivates the default mode network by letting go of the internal dialogue, which moves us from sympathetic (fight or flight state) to parasympathetic mode (relaxed state). I spoke to yogi and therapist Carolyn Cowan on how to achieve a relaxed state and the interconnection between mental health and the gut. Waking up to this way of being brought my mind to what is achievable in our dream state. Dream yoga, or lucid dreaming, is when you become conscious when asleep. Lucid dreaming expert and my sister's adolescent partner in crime Charlie Morley had been dropping off books he's penned on the subject for years. The book that jumped out at me was *Dreaming through Darkness* about how you can heal trauma when you wake up in your dreams. My conversation with Charlie was one that shifted my perception of consciousness and in some ways reality, the power of the human mind and our place within the universe.

Mental health extends far beyond the self. We are, after all, social animals and have thrived in being so. The first lockdown saw the suppliers of

Petersham Nurseries lose their income. We took out the furniture from our restaurant, brought in fridges and laid out boxes of veg to help them set up an order-only farm shop. With the bustling kitchens now quiet as a whisper, using the largest pots I could find I began to cook hundreds of meals for the local food bank. Overnight, families had found themselves below the poverty line. My family home is connected to our restaurant and friends who had moved in to weather the storm also set about helping. In a time of uncertainty, it provided a sense of community. Community is at the heart of *Recipes to Reconnect* and for that reason, my very first conversation was with my friend and explorer Bruce Parry to talk about what he has learnt from indigenous egalitarian cultures who still hold the true essence of community.

I wrote this book with conversation and food at its core, because that is where my own enjoyment and learning have come from. The conversations that lead the narrative are often continuations of dialogue started at the dining room table. Food nourishes our beings. It is the beating pulse of our existence, and my ever-evolving relationship with it has taught me about the environment and human health, as well as connecting me to others. Coming together for a meal holds communities together, be it in the wake of death, to celebrate love or life. Sunday lunch was my family's ritual, marrying the Italian need to feed with Australian warmth. We adopted it from the British and it allowed us to grow roots within this rich culture. A huge part of my own education came from that table and the characters that populated it. A shared meal gives way to thought and feeling – it can be an intimate sharing or a loud and animated affair. It is my anchor to the people I love and those whom I welcome into my home. Dutifully I scribbled down, cleaned up and transcribed the words from each conversation before handing them over to the incredibly talented chefs who believed in this project and lent support through their recipes. Each painted a picture with flavours, fine produce and their own unique approach. This is not a guide to health but an explorative journey, where the reader plays just as much a part in this quest to find out what affects our mood, health and happiness. This book is to encourage an ever-evolving discussion about syncing with the seasons and the food that sustains us. Researching this book has filled me with awe and positivity about all the incredible individuals and movements that are creating change. I implore you to turn away from the onslaught of bad news and actively learn about and contribute to a greener, healthier and happier future.

SPRING

Just as the natural world is being reborn we have our own transformation; we become more energized and outgoing, Spring is a time of year when we align with nature. It is the perfect time of year to think not only about human but also planetary health and how they are woven together with an invisible string. In this chapter, we look at how to keep in balance with the world around us, whilst turning the seasonal produce into mindful dishes.

Living in harmony is an important thread throughout this book and naturalist, philosopher and activist Satish Kumar highlights that we are more intertwined than we think – just as the tulips have grown out of the soil, in some regard so have we. I talk to Ayurvedic author and practitioner Dr Robert Svoboda about keeping our external and internal life in equilibrium and I catch up with my organic-farmer younger brother about equilibrium within farming. Harry is a constant source of inspiration and knowledge and we talk about circular farming, how to bring nature back and how to utilize ancient practices to help our climate and food systems.

In this section, I have paired Satish Kumar with The Petersham, as his ethos has long inspired our family business. Dr Svoboda is linked to Avinash Shashidhara, one of London's best Indian chefs, who grew up surrounded by the Ayurvedic philosophies. Chef Jeremy Lee, an enthusiast for the finest produce and often overlooked offcuts, meets with my brother Harry Boglione.

WE
ARE NATURE

One of my first memories of being truly happy was playing in a giant fig tree; it filled the back garden of my auntie's house and to us children this tree was everything. We created houses, tea stations and lookout points, we breathed in the musky scents of leaves, dust and scraped knees, we watched naughty possums above and chickens below. Held by the thick branches and hidden by the leaves, afternoons were spent blissfully in another world. Nowadays I seek out that feeling, one of being grounded, excited and connected to something greater than myself. I look for it in all corners of the globe, interchanging landscapes filling my horizon: in the frozen sea, white and icy, in freshwater lakes, in hot, sandy dunes and more often in dense, moist, green forests and mountain tops. It is easy to forget that we carry this connection to nature within, that we are nature.

We look to indigenous people to teach us a connective way of being, how they whisper to the forest and ask plants to heal. We are fascinated by something we believe we have lost. Bricks and mortar have risen up between us and mother nature, causing a separation from the ecosystems that keep our lives in balance. The rapid transformation of food systems has had a profound effect on our health, environment and economy. We have a choice, to fall over the tipping point or to take control of our future. On the subject of connectivity, Satish Kumar writes that 'The soil is a metaphor for entire natural systems. If we take care of the soil, the soil will take care of us all.'

Satish explains during our conversation that we are the earth, grown out of it and transformed. We exist within a united tapestry, each thread stitched together; if you tug at one they will all start to unravel. As we experience increasing climate change, this is becoming ever more apparent: whole ecosystems are being wiped out and multiple species are on the decline. Economy and ecology should be able to thrive together. There needs to be a shift in collective consciousness to see nature as a key part of our future. The UK Treasury appointed Sir Partha Dasgupta of the University of Cambridge to write a review exploring 'The Economics of Biodiversity'. In the report, Partha writes that 'Biodiversity is the diversity of life. We will find that the economics of biodiversity is the economics of the entire biosphere. So, when developing the subject, we will keep in mind that we are embedded in nature.'

A conversation with

SATISH KUMAR

I first heard Satish Kumar talk when I was an adolescent, my mind starting to expand and full of questions. Satish was initially inspired by his mother's teaching of nature and Mahatma Gandhi's non-violent action toward people and the planet. He is an Indian-born naturalist, pacifist and activist. At the age of nine, Satish became a Jain monk. Having found 'new consciousness' at the age of eighteen and believing that spirituality should be available to everyone, he ran away. A few years later, he made headlines by walking 8,000 miles for peace, delivering 'peace tea' to world leaders. In 1973, Satish settled in the UK and has been editor for over forty years at *Resurgence & Ecologist* magazine. Satish co-founded the Schumacher College and served as Programme Director until 2010. He has authored many books, including his autobiography *No Destination* and *Soil Soul Society*.

AB One of the inspirations for this project was a sentence in your book *Soil Soul Society*, where you write: 'There is no healing the self, if the earth around us is sick and human communities are suffering.' Can you elaborate on this?

SK As I said, unity of life is fundamental. Out of unity comes diversity. At the time of the Big Bang, there were only gases and then water, and then slowly, slowly, evolution created diversity. We all come from the one source, one origin. Everything is connected. Everything is interdependent. Soil, Soul, Society is the trinity. Soil is the source of life.

Our body is soil transformed. Our clothes are soil transformed. Soil, with the help of rain and the sun, transforms itself into food. So, what we are eating is food, and food is soil. So, food is soil, our body is soil, our clothes are made of soil – the cotton, the wool – everything that we wear is soil – but now nobody talks about soil. We take soil for granted. We all need to touch the soil, feel the soil, protect the soil, build the soil, by putting compost on the soil and trying to stop the soil from eroding, making sure the soil is in good heart.

If the soil is in good heart, everything will be in good heart. If we look after the soil, the soil will look after everything else. So, I put soil as a symbol of all nature, a symbol of all our planet Earth. For me, soil is kind of one word, which is such a beautiful word. It touches my heart, touches my imagination, touches my poetry and everything. Lots of people don't see the internal, they only see the external. Soil is external and soul is internal. So, we also have to have that internal cultivation. We have to cultivate our inner spirit. We have to cultivate our relationship with the soil and our relationship with people. That relationship comes out of love, out of compassion, out of kindness, out of generosity. All of these qualities have to be cultivated and that is cultivating the soul. These are soul qualities. Like soil has qualities, soul has qualities, and those qualities are love, compassion, kindness, generosity and hospitality.

The soul is the bridge between nature and humans. We are the bridge. Through ourselves, we make connections. So, cultivating our soul and looking after ourselves and being happy, being joyful, being blissful, being calm, being relaxed, being happy and healthy, are prerequisites for any activism, social action and for any environmental action. Therefore, I put great importance on soul as being the middle of the trinity.

I put society as a third element because we are not just individuals, we are individuals totally interconnected with the rest of humanity. Therefore, we need to feel that sense of unity with all people and make sure that no one in our society is neglected, isolated, poor, hungry or ill. If they are, then we go and reach out to all those people. We start in our local community – in our village, in our neighbourhood, in our city, in our country. Wherever we are, we start with local but we have a global mind. We have love and care for the whole of humanity, whether you are a communist, capitalist, Christian, Hindu, Muslim, Buddhist, Black or white or man or woman, younger or educated. That common humanity has to be cherished and celebrated with love, care and compassion for all living beings. Soil, soul, society, this is the new trinity for our time. If we can bring these three words together, then we can create a holistic movement.

When I talk about the environmental movement, I always say that there is a natural environment, there is a spiritual environment and there is a social environment. The environment is not only trees, forests, mountains, animals and birds, rivers and oceans. We also have an inner environment, which is the soul environment.

AB Anna Boglione
SK Satish Kumar

There is a spiritual environment and we have to be careful to look after our inner environment as well as the social environment. Social interaction, social relationships, all those things have to be looked after. This way I make the holistic environmental movement – natural environment is soil, the spiritual environment is soul and the social environment is society. These are the three dimensions.

AB   It's really interesting to see the time that we're living in as we've found ourselves in a very peculiar moment. Covid has shown us that we are all human. The virus doesn't discriminate; it infects world leaders, princes, students, our friends. It has shown us that we are one community and has, in turn, somewhat brought us together to remember that we are social animals. We're also realizing we want to be in nature, as well as that we want to look after each other.

I believe that all the planetary systems have to be working in unity with each other and in equilibrium. We can achieve a holistic future, but we need to change our education and look after our environment. This is a large thing to restructure our economy with nature rather than against nature.

SK   Economy and ecology come from the same root. The Greek words are *oikos*, *logos*, *nomos*. 'Oikos' means home, household. 'Logos' means knowledge, study of knowledge. So, knowledge of the household is ecology. Economy is 'nomos', which means management. So, proper management of the household, or Earth household, is economy. At the moment, people think economy is about money, profit, production and consumption. That is a mistaken view of the economy. Economy is the proper management of the Earth's household all together. In our universities and schools, they teach economy without teaching ecology. Before you manage something, you have to know what you are managing.

Learning about ecology means knowing about the household. Ecology and economy are like walking on two legs. They are the two sides

If we look after the soil, the soil will look after everything else.

of the same coin. Without ecology, there cannot be proper economy. Our understanding of the economy has to change to make sure that the whole Earth system functions together and has integrity. Without that integrity, there cannot be economy. This is the problem of our time. We look at nature as a resource for the economy and not nature as a source of life itself.

If we can change this understanding, then there's plenty in nature and there's no shortage of anything. Mahatma Gandhi used to say that there's enough in nature for everybody's needs but not enough for everybody's greed. Nature provides so much. If you plant one seed, you get a thousand fruits back. Nature is a multiplier, but the human economy has created scarcity. This is the big challenge we face today in the world.

AB   Do we need to measure GDP by taking in environmental health, human growth and community and not just focusing it on monetary growth?

SK   Absolutely. We need to change our measurement. Instead of economic growth, we have to think about growth in well-being. A good example is Bhutan. In Bhutan, the government does not measure economic growth, or GDP; they call it GNH – Gross National Happiness. The integrity of communities, families, culture, the relationship between humans and nature, how people are happy, their well-being and health and how good is their community health, all those things are incorporated into that Gross National Happiness measurement. This small country, one of the smallest in the world with only one million people, is setting a good example. The big countries need to learn. European countries, the United States of America, Canada, Japan, China and India, all the big countries, need to shift their measurement of the economy in terms of money and money movement. Health, happiness and community, family and personal well-being are the real measurements and we should focus on them.

THE PETERSHAM COVENT GARDEN

Set within the cobbled streets of Covent Garden, The Petersham steps away from the dirt floors, rickety tables, plants hanging overhead and general bohemian vibe of our flagship restaurant Petersham Nurseries. Abundant with flowers, splashes of colour provided by contemporary art, it reflects the clean lines and elegant style of our family home. The slow-food philosophy of good, clean and fair assures meticulously sourced ingredients, nutrient-rich produce and well-reared meat, often from my brother's farm, Haye Farm Organics. The Petersham creates wonderfully refined menus. As with our other restaurants, the thread that runs through the culinary experience is the unique balance between the best of British produce and Italian culture.

Satish Kumar has been a huge influence within my own narrative, as well as an enthusiastic supporter of our restaurants. With this in mind, I had to open the book with Satish's words, whilst marrying them to the fine and subtle flavours of The Petersham. In these recipes, we wanted to celebrate the beautiful bounty mother nature throws forth and honour Satish's vegetarian heritage. Each recipe is simple yet poignant, as are Satish's words.

Artichokes alla Giudia

Artichokes alla Giudia (Jewish-style artichokes) originates from the Jewish areas of Rome and is typically eaten in the springtime. Artichokes are a thistle, cultivated centuries ago in the Mediterranean for their medicinal properties and one of the true delights of the spring season. Here they are eaten very simply to show off their unique flavour.

4 large globe artichokes
2 lemons (juice one and cut the other into wedges)
vegetable oil, for frying
15–20 sage leaves, stalks removed
salt and pepper

SERVES 4
Cooking time approx. 30 minutes

METHOD
To start, fill a bowl with cold water and add the lemon juice to it – this will stop the artichokes from discolouring. With a sharp knife, remove all but 5cm (2 inches) of the artichoke stalks. Using a vegetable peeler, peel the stalk down, removing the green outer part. Snap off the thicker outer leaves of the artichoke and cut the top off the artichoke, leaving just the yellow base attached to the stem. Remove the furry-looking purple inside leaves of the artichoke and, once the furry interior is cleaned out, place the remaining artichoke in lemon water.

In a deep pan, heat enough oil to cover the artichokes to 200°C (400°F). Drain the artichokes well and carefully add to the oil. Fry till brown and crispy, turning throughout. Remove from the oil and drain well on kitchen paper.

In the same oil, fry the sage leaves till they turn dark green – this should only take 10–15 seconds. Season the artichokes with salt and pepper and serve with a lemon wedge and fried sage.

Asparagus with egg yolk & Parmesan

No other British ingredient seems to have quite the impact of asparagus. Heralding the start of spring and only around for a short time, it is incredibly versatile and can be boiled, fried or just shaved and eaten raw in a salad. In this recipe, we serve it with its natural partner, the egg!

18 free-range eggs
24–36 asparagus spears, 4–6 per person, ends trimmed
150ml (¼ pint) olive oil
15g (½oz) thyme
15g (½oz) basil
15g (½oz) dill
150g (5½oz) Parmesan cheese, grated
salt and pepper

TO SERVE
1 teaspoon olive oil
6 nasturtium leaves
15 chervil sprigs
18 basil leaves, torn
15 dill sprigs

SERVES 6
Cooking time 30 minutes, plus overnight freezing and marinating

METHOD
Place the whole eggs in the freezer overnight.

In salted water, boil the asparagus for 2 minutes, then chill in ice water. Once cooled, drain the asparagus well and dab on kitchen paper to fully dry. Place the asparagus in a container with the 150ml (¼ pint) of olive oil and herbs and leave to marinate for as long as possible, but 4–12 hours would be ideal.

The next day, remove the eggs from the freezer and allow them to thaw naturally. When fully defrosted, crack the eggs and separate the yolk and whites (reserve the whites for other dishes). Mix the yolks with salt and pepper and store in an airtight container or piping bag until needed.

Preheat the oven to 175°C (350°F), Gas Mark 4. Spread the Parmesan on parchment paper in roughly 5cm (2inch) circles to create 12 Parmesan crisps. Bake just until the cheese has fully melted. Allow to cool before adding to the dish.

Heat the teaspoon of olive oil in a shallow pan, then remove the asparagus from the herbs and oil, season with salt and pepper and add to the pan. Fry until lightly coloured, then remove and drain on kitchen paper. Arrange on plates, spoon or pipe on a few coin-sized dollops of egg yolk and garnish with the Parmesan crisps, nasturtium leaves, chervil, basil and dill.

Cacio e pepe bottoni

Cacio e pepe is an ancient Roman dish, fit for shepherds, as the legend goes. They would take dried pasta, pecorino cheese and an abundance of black pepper with them whilst walking their herds through the hills, creating this much loved and classic dish. Here we have flipped it slightly, serving the rich cheese and pepper filling inside our own ravioli 'buttons'.

PASTA DOUGH
105g (3¾oz) 00 flour, plus extra
    for dusting
45g (1½oz) semolina
90g (3½oz) egg yolks (from
    about 5 eggs)
3 Free-range eggs
30ml (1fl oz) olive oil
a large pinch of salt

CHEESE FILLING
75g (2½oz) butter
75g (2½oz) plain (all-purpose) flour
1.5 litres (2½ pints) whole milk
300g (10½oz) pecorino cheese, grated
75g (2½oz) white bread, crusts removed
½ teaspoon salt
½ teaspoon black pepper

SAUCE
50g butter
30ml (1fl oz) olive oil
50ml (1¾fl oz) water
pinch salt
pinch black pepper

TO SERVE
pecorino cheese, grated

SERVES 6
Cooking time 1 hour, plus 30 minutes chilling

METHOD
Mix all the dough ingredients together slowly in a kitchen mixer, allowing the flour to be incorporated. Remove from the mixer and knead on the countertop for 5 minutes until the dough feels smooth and springy with no cracks. Wrap in clingfilm (plastic wrap) and rest in the fridge for at least 30 minutes.

To make the filling, melt the butter in a pan, add the flour and mix well. Slowly add the milk, continually whisking whilst adding the cheese, and mix till smooth. Roughly break up the bread and fold it into the mix, before blending the mix either with a hand blender or kitchen mixer. Make sure the mix is smooth, adjust the seasoning if needed and leave to cool before placing it in a piping bag.

Sprinkle the countertop with flour and roll out the pasta dough. Using a pasta machine, roll the dough out, starting with the thickest setting and gradually running the dough through thinner settings until the dough is at its thinnest and is smooth, making sure there are no bumps. Once rolled out, lay the pasta sheet down on a flat surface, roughly marking where the halfway point is. On one half, add dollops of the cheese mix, making sure to spread them out evenly in rows, with 2 fingers of space between them. Once you are finished laying out the filling, brush the second half with water and gently fold over the top half of the pasta to cover the filling. Gently press down around the filling, making sure both sheets of pasta meet. Using a ring or cutter just larger than your dollops of filling, cut the shapes out and store them on a floured tray until ready to cook.

Fill a deep pan with water and bring to the boil with a handful of salt. Cook the bottoni for 3 minutes then drain. In a separate pan make the sauce: melt the butter, olive oil, water, salt and pepper together to create an emulsion. Take the sauce off the heat and add the pasta, tossing till well coated before plating. Serve with grated pecorino.

Catalan salad 'cestino di croccante' with spring vegetables & herbs

Catalan salad is a fresh and vibrant staple in Sardinia, traditionally served with lobster and seafood. Here we have served it with just the vegetables in a crispy basket as the mixture of textures and flavour really stands up on its own. Why not get your hands in the soil and grow your own ingredients, or if you have a vegetable garden already, try using what's there.

12 baby carrots, trimmed
2 celery sticks, trimmed
1 courgette (zucchini), trimmed
12 radishes, trimmed
12 cherry tomatoes
herbs (basil, cress, chervil, celery, 2g (⅓oz) total)
6 asparagus spears, trimmed
60g (2¼oz) fresh peas
60g (2¼oz) fresh broad beans
4 spring onions (scallions), trimmed
12 baby spinach leaves
handful of edible flowers
salt and pepper

LEMON DRESSING
40ml (1½fl oz) lemon juice
4 teaspoons Chardonnay vinegar
4 teaspoons water
1 teaspoon sugar
½ teaspoon Dijon mustard
1 teaspoon salt
75ml (2½fl oz) olive oil
75ml (2½fl oz) vegetable oil

MISO EMULSION
1 clove garlic
2cm (½ inch) fresh root ginger, peeled
2 tablespoons rice wine vinegar
30g (1oz) miso paste
60g (2¼oz) egg yolks (from about 4 eggs)
100ml (3½fl oz) water
20g (¾oz) honey
1 teaspoon salt
210ml (7½fl oz) vegetable oil

CRISPY PASTRY BASKETS
50g (1¾oz) honey
100g (3½oz) butter
4 teaspoons lemon juice
spring-roll pastry

SERVES 6
Cooking time 40 minutes

METHOD
Preheat the oven to 170°C (325°F), Gas Mark 3.

Peel the baby carrots and thinly slice them into rounds (you can use a mandoline to make it easier). Do the same with the celery. Thinly slice the courgette and radishes into rounds and halve the cherry tomatoes. Pick the herbs and store in cold water until needed. With a vegetable peeler, peel the asparagus spears into thin strips from tip to tip and place in water.

Shell the peas and broad beans and cook in salted water for 2 minutes. Thinly slice the spring onion and store in cold water until needed.

For the dressing, place all the ingredients in a bowl and, with a hand blender, blend together.

Take all the ingredients for the miso emulsion except for the oil, place in a blender and mix for 1 minute. Slowly add the oil once the other ingredients are mixed together.

To make the pastry baskets, warm the honey and butter together and, when the two are incorporated, add the lemon juice. Brush one sheet of spring-roll pastry lightly with the honey and butter mix and then fold the sheet in half, creating a double layer. Brush the top of the double layer with more honey-butter mixture and shape into a 10cm (4 inch) tart case or a mould of your choosing. Repeat to make 6 baskets. Bake with no fan for 5 minutes.

To serve, coat the vegetables, herbs, spinach leaves and flowers lightly with the lemon dressing and season with salt and pepper. Arrange in the pastry basket and top with the miso emulsion.

Strawberry vegan pavlova

A fantastic dish, unintentionally vegan and bursting full of flavour.
The pavlova uses the leftover and often discarded water from
tinned chickpeas in place of egg whites, which provides a very unique
and nutty flavour.

PAVLOVA
120g (4oz) aquafaba (chickpea water from
   approx. 1 tin)
100g (3½oz) caster (superfine) sugar
1 teaspoon vinegar

STRAWBERRY SORBET
100g (3½oz) strawberry purée
40g (1½oz) caster (superfine) sugar
55ml (2fl oz) water

STRAWBERRY SALAD
250g (9oz) strawberries
2 teaspoons lemon juice
50g (1¾oz) icing (confectioner's) sugar

SERVES 6
Cooking time approx. 3 hours

METHOD
Preheat the oven to 90°C (200°F), Gas Mark 1/4.

Add all the pavlova ingredients into a bowl and either by hand or
with an electric whisk, beat together until soft peaks form. Use
a large spoon to dollop onto a baking tray into 6 circles roughly
10cm (4 inches) wide and 2cm (¾ inch) deep and dry in the oven
for 2 hours.

To make the sorbet, boil all the ingredients together for 5 minutes
– it should have a thick, jammy consistency. Allow to cool before
churning in an ice-cream machine. This should take roughly
5–10 minutes.

Halve the strawberries and toss together with the lemon juice
and sugar. Leave to macerate for 10 minutes.

Place the meringue circles on 6 plates and put the strawberry
salad on top. Finish with a scoop of sorbet.

HOLISTIC

In Sanskrit, ayur means life and veda means knowledge, so Ayurveda translates as 'knowledge of life'. This ancient knowledge is based on what we now recognize as modern concepts. It is said that Vedic yogis meditated on the earth's frequencies, tuning into the quantum field to understand the frequencies of the physical body, earth and soul. Our bodies, both physical and non-physical, are influenced by these energy fields. Prana is the manifested energy of the entire universe; it translates from Sanskrit as 'life force energy' or 'breath of life'. Prana flows in currents around the body and not only governs functions such as respiration, circulation and digestion but also coordinates our breath, senses and mind. Paul Dugliss, Academic Dean and Director of New World Ayurveda School, says 'everything derives from a unified field of energy and intelligence. One field of consciousness.'

Ayurvedic philosophy is based on the Samkhya school of thought that believes all things are created from the inherent force of nature known as prakṛti and illuminated by the primary force of consciousness, puruṣa. All life observed in manifest reality arises from the dance between puruṣa and prakṛti, consciousness and creation. Ayurveda looks at all aspects of life, including diet, environment, movement and even virtual spaces you might exist within. Without understanding a person as a whole, with all their influences and inputs, you will not be able to manage their illness or wellness. By focusing on the connectivity of all inputs, rather than on the singular, harmony can be be reinstated within the self.

LIVING

In order to maintain balance it is important to understand one's constitution; by understanding yourself both energetically and physically you empower yourself on a path to wellness. We are each made up of the five elements: air, space, fire, water and earth. There are three elemental combinations, or doshas: Kappa, Pita and Vatta. Kappa's primary element combination is water and earth, Pit's is water and fire and Vatta's is air and space. Each of our constitutions can be aligned with one or more of the Vedic doshas. Once you have a basic understanding of your dosha you can use it as a guide, alongside herbs, movement, meditation, breathwork and daily ritual.

Ayurveda looks at all aspects of life for healing and medicine. It is not about 'fixing' yourself but instead seeing the whole journey and life itself as sacred. By following a system that puts life in the centre we align ourselves with a life path that helps us thrive as individuals, supporting and contributing to the wider ecosystems.

A conversation with

DR ROBERT SVOBODA

The rain is falling above my head as Dr Robert Svoboda appears on my screen. Speaking to me from Texas having spent lockdown with his sister, his energy is kind and that of an old friend. We easily slip into conversation as he dances from topic to topic, his wisdom and knowledge apparent. Robert was the first western person to receive his Bachelor of Ayurvedic Medicine and Surgery (Ayurvedacharya) from the University of Poona in 1980. Whilst studying under his mentor Aghori Vimalananda, he also learnt about other forms of classical Indian lore, such as Jyotish and Tantra. Robert has written over a dozen books on the sciences of India and travels the globe lecturing on the subjects that he has long studied.

AB What would you say to people suffering with, or giving into, cravings? Our minds sometimes trick us into wanting sugary, fatty and unhealthy food. The more we listen to our bodies, the more we slow things down, does it become easier to ignore the cravings or do they just dissipate?

RS In my opinion and in my experience, cravings are able to work on you more efficiently when you are paying less attention to them. However, when you pay too much attention to them, it also has an undesirable effect. It's useful to acknowledge that your body craves things – things that it either personally desires or that it has been trained to desire.

AB Emotionally desire?

RS Yes, very commonly emotionally desired. It's good to remember that sugar will always be appreciated by a human organism, because it provides immediate fuel to the system. It's also good to remember that up until a couple of hundred years ago, sugar was a luxury item. Almost nobody ate it, because it was not available. Fat is something that humans will always be attracted by because it provides twice as many calories as carbohydrates and it is a good way for the body to store energy. So there's always going to be an appreciation for fat.

Initially, if you taste something that is intensely sweet, your body will recognize it as being too sweet. But if you build up a tolerance to it there will come a point where you cannot recognize that level of sweetness, unless something is dramatically too sweet. That's the point that the modern world has reached. When you have a can of Coca-Cola that contains ten teaspoons of sugar, you're not even paying attention to the taste anymore. Your body has simply become habituated to a certain amount of sugar being delivered to it all the time.

For those people who are able to exert a reasonable degree of self-control, it is often useful to actively limit yourself. The Ayurveda texts suggest, whatever your quantum of addiction, cut it in half first. Cut the dependency in half, become stabilized at that, then cut it in half again, and become

AB Anna Boglione
RS Robert Svoboda

stabilized at that. Usually after you have reduced your intake of something dramatically, then stopped it completely for a while, you will find that it does not taste as good as it used to when you come back to it. Your body has reorganized the way that it tastes.

AB It's amazing how much your taste buds can change. I remember doing this diet where I cut out all sugars, even things like potatoes, that turn to starch and then to glucose. After a few months of this, my taste buds found even natural sugars overpowering. Everything shifts and you start to truly taste things again.

RS It is only in the past sixty years that the majority of people in the world have had food available to them twenty-four hours a day, and have had pretty much any amount and variety that they want. Now everybody can consume all the time. I personally like the intermittent fasting approach of only eating during an eight-hour period, though I admit sometimes it stretches to nine or ten hours. I can tell that my body likes this too. I get up in the morning and only have water for at least the first two or three hours, then I will have fat, which I usually put in coffee (nowadays I drink decaf because I don't want the blood sugar destabilizing effect that caffeine provides). When you do eat something, your body will not be demanding very strong flavours, very intense tastes and very large amounts of those things that are going to cause your blood sugar to spike up and then crash later on.

AB Is that why people feel lethargic after eating? They eat things that are too complicated, and too much of them?

We're isolated from the natural world, but... we're also isolated from one another.

RS A large amount of that is to do with blood sugar, though not immediately after eating. If a person feels lethargic immediately after eating, most commonly that is because they've eaten too much and all the blood in the body has gone to the digestive tract and there's simply not enough to maintain their cerebral function. Your brain, your eyes and your skin don't store glucose; they have to have an ongoing and continuous supply of it. So if the blood is not getting there you're not getting enough glucose and you're going to feel lethargic. Your nervous system will always prioritize your reptile brain (controlling our vital functions) over your cognitive function, because cognitive function is not essential while the reptile brain is essential.

AB Ayurveda is very much about food and using food to keep that equilibrium. Does it work similarly to western herbalism by using a lot of different and often common plants?

RS Yes, certainly, and particularly regarding what we think of as spices. There was a very well-known Ayurvedic physician in South India, who has just recently passed away, who would only treat people with common spices because he felt what's the use in telling your patient to go get something really obscure and expensive when that's going to make them more stressed. Ease is what we're looking for.

AB Even British rosemary, thyme, parsley and everything that we've got in our herb gardens have medicinal properties. You would imagine that our ancestors incorporated all those different herbs for medicinal reasons.

RS When I went to India first, which was almost fifty years ago, it was still very common for mothers to keep a certain variety of herbs to use for first-aid on their children. Sadly, as is the case in so many other parts of the world, that has become less common because it is much too easy to pay attention to the advertisements and the pharmaceutical industry and simply go somewhere and obtain a pill or a decoction of some kind.

AB I'm a strong believer in the power of plants. There are three sections of Ayurveda: the body, the mind and then your whole world. How does one keep the mind in balance and equilibrium?

RS An important part of that is keeping your Prana, your life force, in balance. This means living a 'regular' life. If you look at the so-called blue zones around the world where people frequently live to an advanced age, you will find that these people live lives that many of us in the modern world would describe as boring. They get up at the same time, they go to sleep at the same time, they mostly do the same things daily, they have strong connections to their family and friends, and they maintain those connections. They're integral parts of their societies and communities; they provide assistance when assistance is required of them and they accept assistance when they need it. That's the way that humans have traditionally always lived, but unfortunately all around the world now there is a profoundly greater degree of isolation among individuals. It's bad enough that we're isolated from the natural world, but in addition to that, we're also isolated from one another. The pandemic has just made this all the more obvious. Even before then, many people were simply living lives in which they didn't interact with one another. Humans are an extraordinarily social species. After we're born we're completely helpless for many years, which is not the case with most animals. We are extraordinarily dependent on one another. One of the effects of this need not being taken care of is that people are finding it easier than ever to get involved in bizarre cults and conspiracy theories, because it provides a sense of belonging.

An important part of being healthy, once your organism is nice and balanced, is to find some way to make your mind, your emotions, your soul and your spirit feel satisfied. Maybe it's by doing some kind of work that is satisfying and rewarding, maybe it's by being available and helpful to your family or to other members of your society or maybe it is by investing a good amount of time in communing with the spirit.

AVINASH SHASHIDHARA

After a decade at the River Café, Avinash Shashidhara left to start a new venture, likening his departure to leaving home all over again. This time it was to set out on a journey to take the melting pot of traditional Indian flavours into the realm of fine dining – realized in the form of restaurant Pahli Hill. Located in the heart of Fitzrovia, London, the menus are inspired by the diverse and vibrant Indian culinary culture coupled with British seasonal produce, each dish celebrating this electric combination. I was blown away the first time I ate at Pahli Hill and every time since.

When I asked Avi what he drew from my conversation with Robert, he told me that the ethos of Ayurveda, how people eat and live, is deeply embedded in his culture and Indian cooking. 'Growing up in India, you naturally eat through the seasons,' Avi says. He explained that he looked at what was in season in the UK and picked the things that resonated with the science of Ayurveda. He chose asparagus as a superfood; hemp seeds, which are a very common ingredient in southern India; and nettles, which have medicinal properties and are found in the Himalayan belt. Avi also remembered his mother's cooking and how she would make him teas for stomach pains and use pinches of spices in dishes as remedies.

Chickpea & fennel salad

This dish can be made with fresh peas, broad beans (when they are in season) or even with black or white chickpeas. It is usually served on its own, as a snack, or with flatbread made with either rice flour or wholemeal flour. My version makes a crisp salad with shaved fennel bulb and curry leaves.

1kg (2lb 4oz) dried chickpeas, soaked overnight in cold water
50ml (2fl oz) sunflower/coconut oil
1 tablespoon black mustard seeds
6 sprigs of curry leaves, plus some fried in oil till crispy for garnish
8 green chillies, finely sliced
6 red onions or shallots , finely sliced
a bunch of fresh coriander (cilantro) leaves, chopped
1 fresh coconut, grated
juice of 3 lemons, plus extra for the fennel
2 fennel bulbs
salt

SERVES 6
Cooking time 1 hour

METHOD
Cook the chickpeas for about 30 minutes in fresh water or until soft. Season with salt and set aside, retaining just enough water to cover the chickpeas.

Heat the oil in a pan. Just before the oil starts to smoke, turn the heat down and add the mustard seeds. When the mustard seeds start to crackle, add the curry leaves, green chillies and onions and fry until the onions are translucent.

Add the chickpeas and their water and continue to cook, stirring occasionally, until the chickpeas absorb the flavours and the water reduces by half. This should take about 15 minutes.

Add half the coriander leaves and the coconut, then cook until most of the water has evaporated. Remove the pan from the heat and add the rest of the coriander and the lemon juice.

Thinly shave the fennel on a mandoline or slice with a knife and dress it with lemon juice and salt to taste. Serve with the chickpeas, either warm or at room temperature, and crispy fried curry leaves.

Rasam

Rasam is one of my all-time favourites, and it showcases Karnataka's black pepper like no other dish. A hot-and-sour broth made using toor dal, tomato and tamarind as a base, it is flavoured with a good handful of toasted and crushed black pepper, cumin and fresh coriander. It is best served on its own or with a little steamed rice.

100g (3½oz) yellow toor dal or split red lentils
1 teaspoon oil, plus 1 tablespoon for tempering
¼ teaspoon turmeric
2 plum tomatoes, cut into quarters
1 teaspoon black peppercorns, toasted
   and crushed
1 teaspoon cumin seeds, toasted and crushed
1 tablespoon tamarind pulp
½ bunch of fresh coriander (cilantro)
½ teaspoon mustard seeds
2 garlic cloves, crushed
1 dried red chilli
1 sprig of curry leaves
a pinch of asafoetida (optional)
salt

SERVES 4
Cooking time 1 hour

METHOD
Rinse the lentils and place in a saucepan with enough water to cover them, then add the teaspoon of oil, turmeric and tomatoes and bring to the boil. Skim the froth and simmer for about 30 minutes until the lentils and tomatoes are almost mushy, then gradually add water to the mixture to bring it to a broth consistency.

Add the black peppercorns, cumin, tamarind pulp, coriander leaves and a pinch of salt. Continue to cook for a further 10 minutes to extract all the flavours from the spices.

In another shallow pan, heat the tablespoon of oil and crackle the mustard seeds, garlic, chilli, curry leaves and asafoetida. Sauté for 30 seconds, being careful not to burn any of the ingredients as the oil can get hot very quickly. Add the temper to the broth and serve with rice or on its own.

Saag paneer with spring nettles & spinach

Paneer, or cottage cheese, is made by bringing either cow's milk or buffalo milk up to the boil and curdling it by adding some lemon juice or vinegar. It is then wrapped in muslin and suspended, or pressed under a weight, in order to get ride of the water. It can be shaped into blocks or crumbled and added to dishes. You can make this at home or it's easy to find it in most British supermarkets.

Saag means 'greens' in Hindi and a variety of greens can be used to make this dish. I like to use nettles in the spring as they are the first of the greens to grow after a cold winter, when the soil has had the time to rejuvenate and is rich in minerals. Nettles are rich in nutrients and have an amazing earthy flavour. The addition of creamy, bouncy paneer makes this especially delicious.

1kg (2lb 4oz) nettles
500g (1lb 2oz) spinach
2–3 green chillies
2 tablespoons sunflower oil
1 teaspoon cumin seeds
3 garlic cloves, finely chopped
1 teaspoon coriander seeds, toasted and crushed
250g (9oz) paneer, cut into 2–3cm (1 inch) cubes
salt and pepper

SERVES 4
Cooking time 30 minutes

METHOD
Pick and wash the nettles and spinach then cook in boiling salted water for about 5 minutes until they are soft. Make sure you don't overcook them or they will lose all their nutrients and flavour.

Allow the greens to cool, then gently squeeze out the excess water and transfer to a food processor along with the green chillies, with their stalks removed. Grind to a smooth paste and set aside.

Heat the oil in a pan, then add the cumin seeds and heat until they crackle, followed by the garlic. Sauté till golden brown, then add the nettle and spinach purée and heat until bubbling. Adjust the seasoning. Add the coriander seeds and the paneer and cook for a further 2–3 minutes. Serve with flatbread or rice.

Asparagus with bhangjeera chutney (hemp seed chutney)

Hemp seeds, or bhangjeera, are a common ingredient used in the Himalayan regions of India, where they grow in the wild. They can be easily sourced here in the UK and are highly nutritious. They are usually considered a superfood and are used to make a chutney, eaten with breads or other vegetables. It is like an Indian version of a pesto — creamy, nutty and fresh. Perfect for spring! I serve the chutney with another popular super ingredient: asparagus.

a bunch of asparagus, trimmed and
    blanched in salted water
oil, for brushing
sea salt

FOR THE CHUTNEY
200g (7oz) shelled hemp seeds, lightly
    toasted, plus 25g (1oz) for garnish
20g (¾oz) cumin seeds, lightly toasted
3 green chillies
20g (¾oz) fresh coriander (cilantro),
    roughly chopped
20g (¾oz) mint, plus a few leaves
    for garnish
juice of 1 lemon
salt

SERVES 2
Cooking time 30 minutes

METHOD
Add all the chutney ingredients to a blender with a little water and grind to a fine paste, adjusting the seasoning and consistency to taste. Store in an airtight container in the fridge for up to a week.

Brush the asparagus with a bit of oil and season with sea salt. Grill on a griddle pan or barbecue until the spears have a nice char. Transfer them onto a plate, drizzle over some of the chutney and serve with a sprinkling of toasted hemp seeds and chopped fresh mint leaves.

Carrot halwa

Carrot halwa is a dish made all over India but with different variations. It can be rich, sweet and milky, but my version is almost like a carrot cake, with a few sultanas, pistachios and cardamom – not too sweet yet unctuous and moreish. Use red carrots if you can find them, for a better colour and natural sweetness.

500g (1lb 2oz) carrots, peeled and grated
500ml (18fl oz) whole milk
100g (3½oz) caster (superfine) sugar
6 cardamom pods, husks removed and
    discarded
    and seeds crushed
a handful of raisins or sultanas
25g (1oz) ghee

TO SERVE
vanilla ice cream
a handful of pistachios, roughly chopped

SERVES 4
Cooking time 1 hour

METHOD
Put the grated carrot and milk in a heavy-based pan over a high heat and bring to the boil. Turn the heat down to medium and simmer, stirring regularly, until most of the liquid has evaporated. This will take about 30 minutes.

Add the sugar, cardamom seeds and sultanas and continue cooking, stirring so the mixture doesn't catch. Finally, add the ghee and cook until the colour is deep orange and unctuous. Allow it to cool slightly before serving with vanilla ice cream and chopped pistachio nuts.

NOTE
The sweetness of carrots can vary, so it is best to adjust the sugar according to taste.

SOIL

When politicians were looking to fill the hunger gap left in the wake of the World Wars, the 'green revolution' seemed to be the answer. The introduction of monocultures and campaigns, such as 'Dig for Victory', during and after World War II, saw a shift in the British landscape. Over-tillage plus the introduction of synthetic herbicides in the early 1970s turned our soil from being porous to dense, water runoff and flooding became more common and water-soluble glyphosate found its way into the circle of life, even showing up in mothers' breast milk. There are many papers out there debating the influence of these chemicals on human health: some link them to cancer, others to autoimmune diseases. The microbes in the soil have also been impacted by the change in farming systems and without microbes, the food produced becomes low in nutrients, as it is the micro-organisms that help the plant's root structure absorb these elements.

We are born out of the soil, fed and nourished from it, with all life on this planet sustained by the micro-organisms that keep it healthy. Even the word 'human' derives from the Latin root humus, which means 'soil' and is still used today. For me, the soil is the root to both planetary and human health. In recent years though, it has been pushed into focus for the opposite reasons, such as environmental erosion, loss of topsoil, landslides and desertification.

Seventy years on from the so-called 'green revolution', the issues are ever more apparent. Biodiversity has suffered, the climate is in trouble and our food systems are under serious threat. Julian Cribb, in his book *The Coming Famine: The Global Food Crisis and What We Can Do to Avoid It*, highlights that in the first decade of this century, out of 1.5 billion hectares of global farmland, a quarter was suffering the effects of serious degradation. Desertification and soil degradation can be caused by many things, such as the removal of trees, altered states of ecosystems, bad land management and grazing practices, over-tillage, monoculture farming, global warming and the use of chemicals, pesticides and herbicides. As the philosophy of thought shifts to one that looks at all systems as a whole, the UK government announced the relaxation of legislation around genetically altered crops within the UK (not the direction we want to go in). Hope is still on the horizon as we stand on the edge of another revolution – regenerative agriculture – this time working with nature and not against it. Regen Ag, as it's fondly known, is a holistic form of land/farm management. It looks at reversing desertification, capturing carbon and restoring topsoil, ecosystems, biodiversity and water structures, with a key focus on producing high-quality, nutrient-dense food.

A conversation with

HARRY BOGLIONE

From a very young age, my brother was either slipping through the garden gate and into the plant nurseries or he'd take himself down the lane to the local farmer, collecting horticultural and animal husbandry knowledge as he went. Incredibly dyslexic, but with a brain like a sponge, he taught himself land management and regenerative farming, among other things. When we assembled his first polytunnel together, he had a YouTube tutorial on his phone in one hand. Harry is like that. If he wants to know or learn something, he will find a way that works for him.

Harry runs his 90-acre organic farm, Haye Farm, in Devon, plus an additional 400 acres of surrounding land with his partner and mother of his two children, Emily Perry. Together they farm in a way that increases biodiversity and nurtures ecosystems for future generations. Harry works closely with the Landworkers' Alliance and campaigns for animal welfare, appearing in the documentary series *Farms Not Factories*. He is an advocate for Slow Food and is on a mission to create a network of regenerative businesses.

AB Anna Boglione
HB Harry Boglione

AB We've become conditioned to think certain produce and practices are better, and I think a lot of that conditioning comes from what we see in the supermarket aisles.

HB Conditioning comes from so many places. Farmers weren't very keen to take on chemical agriculture, but chemical companies partnered up with the governments, subsidizing people to do it in the name of feeding the nation. This happened in the wake of World War II and during ongoing rationing to encourage farmers to intensify and focus on becoming a single producer. So, you were either a chicken farmer or a cattle farmer because it was said to be a lot more efficient to just have one function. It might be more efficient in a mechanized way, but you end up with loads of negative externalities, whereas in a mixed farming system there's no such thing as a negative externality.

AB Because everything feeds into something else?

HB The cow and chicken poo make the grass grow. We might plough the field the chickens were in and grow a cereal crop because that field will be very high in the nutrients that you need to grow cereals. Then you've got a cereal crop and you can feed that to the pigs. It's a self-sustaining cycle. If you decide I'm just going to be a pig farmer, you end up buying in loads of grain and taking nutrients away from the cereal farms. So, the cereal farms replace their nutrients by buying in chemical fertilizers, but the chemical fertilizers do not contain a full spectrum of what the plants need, so they produce lower-quality cereals. Then the pig farmer ends up with loads of pig manure that they don't know what to do with. In places in America, they have massive lagoons full of it. They spray it into the air to try and get rid of it, and it makes people sick.

AB How would you normally get rid of animal waste?

HB We have pigs and we keep them in fields and move them every couple of weeks. So, the waste stays in the field and fertilzes the following crop. It's a positive thing. The pigs are like giant worms and the manure is amazing for the soil.

Carbon is not our enemy

AB   Do you think that monoculture farming in the next few decades might come to an end?

HB   Well, it's going to have to change. We need to have more crop rotations and stuff to build soil. We might deplete all the topsoil, but the beautiful thing is then a regenerative farmer can take the farm on and fix it. You can build topsoil, as well as ruin it. There need to be laws in place to not allow the continuous degradation of soil.

AB   Also, if we stopped the degradation of our soil, then we'd also start to fix climate issues. A huge amount of this planet is covered by farms, where we could be locking in more carbon, instead of losing carbon from mismanaged land. If we regenerated soil, if we did cover crops, if we didn't leave the soil to be taken by the elements, we'd be locking in carbon.

HB   A huge percentage of the Earth's surface is a human-created desert, classified as degraded, unproductive land. When I hear a discussion about global warming and climate change, my biggest frustration is that people are just on the wrong trajectory. Carbon is not our enemy, you know, carbon is the most incredible life-giving element in the world. Without carbon, everything's fucked. The focus is always on carbon production and increasing carbon in the atmosphere as a result of burning fossil fuels. In my opinion, global warming is a symptom of degradation. It's not because of the Industrial Revolution. It's not because of diesel-burning cars. It's because we've removed the planet's ability to absorb and cycle carbon. You see it across the board. For example, we've got loads of deforested land that's been farmed very badly and then abandoned. For me, the way to fix climate change is to regenerate these desertified areas, which is very, very doable.

In the tropics, around the equator, you can turn a degraded bit of land into an amazing, fruitful, plentiful forest, with water systems that function, in a matter of years. So, instead of trying to draw carbon out of the atmosphere by using machines or inventing ways of turning it into stone or pumping it into the ground, we should be regenerating these degraded areas. As soon as you restart the carbon cycle and have plants growing, you reinvigorate the water cycle, you rebuild the water tables, you get the streams flowing again and you bring wildlife back into the rivers.

We need to reapply our ability to be engineers and become ecosystem engineers, recreating all the ecosystems that we've destroyed. We need to turn all the deserts back into forests. We need to turn all the dried-out rivers back into flowing rivers. We need to replant all the mangroves we've removed. Nature's amazing; it wants to do these things. Replanting the mangroves doesn't actually involve planting any mangroves; it involves getting in there with a digger and making sure the water flows naturally. If we can modify the land in a way that allows those areas to be waterlogged, the mangroves will float in with the tide, the seeds will plant themselves and they'll grow and become a mangrove.

There are a lot of amazing people, like Gabe Brown, who talks about soil regeneration and Geoff Lawton, with the whole permaculture thing.

AB   It's about working with nature, not against it, which is what we're talking about with regenerative farming. It's also about looking at all the systems as one rather than as individuals. This mindset is so deeply entrenched in us to look at things in a singular manner. We need to look at all the systems in a holistic manner rather than pinpointing and isolating them. So, if we start working towards fixing the root cause, then it will have this knock-on butterfly effect.

HB   It has to be holistic, but I think also we need to understand that humans have a great arrogance about them. We're doing all of this conservation to try and save this one bird or this one frog, when in reality our approach should be, 'We need to make this ecosystem work how it wants to work, and then see what happens.'

AB   What's your opinion on veganism?

HB   Umm, I think veganism is necessary in the system we live in, just because we've got too much of an emphasis on meat-eating and the way that meat is produced in a shocking way. However, I think if everybody turned vegan and we didn't have a demand for meat, it would be cataclysmic to the ecosystems we create and live in. It is very, very, very important that every area of land on this planet sustains grazing animals and the best way to achieve that is by eating the grazing animals. Meat farming has got to be integrated with other methods of farming and land management in order to sequester carbon and get carbon cycles going. The grass has to be turned into nice cow poo or deer poo or whatever poo, which then gets absorbed by the microbiology and feeds the dung beetles and starts another food chain feeding the flies, which then feed the birds and it starts another food cycle. The whole planet would turn to desert if we took grazing animals away because the grasses wouldn't break down. They would just stand there and they wouldn't be flattened by an animal or eaten by an animal, so they wouldn't decompose properly. The only way of it decomposing is ionisation, which is essentially a kind of breaking down back into the atmosphere, and none of that carbon goes into the soil. It's fundamentally important that we have well-managed land with grazing animals on it.

AB   This does not mean factory farming, which does nothing for the planet. People need to make the right choices and know where their meat comes from – more of a farm-to-table approach.

JEREMY LEE

When I pick Jeremy up from the station, he is charismatic and full of charm, ready for a weekend cooking at Haye Farm. Just coming out of winter lockdown, this is the first shoot for the book. We haven't seen too many people and it is a pleasure to have a life-force like Jeremy around. Jeremy is Chef Proprietor at Quo Vadis, an astonishing, old building in the heart of Soho, soon to celebrate its hundredth birthday. Jeremy has appeared on and judged many cooking shows, was a finalist in the second series of *Great British Menu* and is a cookery writer.

With my brother's children in tow, Jeremy is quick to get to work, foraging for our supper and these warm, hearty recipes. With a procession of children, dogs and the occasional cat, Jeremy delves in and out of polytunnels collecting greens, climbing into the back of trailers to snatch handfuls of ancient rye and dipping into the butchery, reappearing with an ox heart twice the size of my head. He is ready to cook. Jeremy has a love of good produce and is excited by his bounty. Being a great supporter of the nose-to-tail ethos, he is particularly thrilled with the magnificent heart.

Ox heart with chickweed salad & wild garlic salsa verde

One of the great things about going to a farm that has its own abattoir is having access to an extraordinary range of the freshest offal and choice of cuts of different meats, which have all roamed free on the farm. Ox heart, as with much offal, is often very difficult to find, particularly in spanking-fresh condition, which is vital. Being a very lean muscle, it can be cooked rare, like liver, and sliced thin, as in the following dish. I chanced upon these lovely leaves as we walked through the farm. However, should these prove elusive consider mustard leaves, watercress, rocket and escarole as alternatives.

SALSA VERDE
a handful of wild garlic
1 tablespoon capers
2 tablespoons apple cider vinegar
5 tablespoons olive oil
1 teaspoon Dijon mustard
a large pinch of black pepper
salt

OX HEART
750g (1lb 10oz) ox heart
½ tsp salt
½ tsp black pepper
4 tablespoons olive oil, plus extra to serve
4 thyme sprigs
1 tablespoon cider vinegar

SALAD
a small handful each of giant goosefoot,
    chickweed, claytonia, purslane, radish
    flowers, bunched spinach

SERVES 6
Cooking time approx. 50 mins
(if the butcher prepares the heart)

METHOD
For the salsa verde, coarsely chop the wild garlic then add the capers. Chop finely together, place them in a bowl, add the remaining ingredients and mix well. Taste for seasoning.

Trim the ox heart of fat, sinew, valves and membrane until only the lean meat remains (you can ask the butcher to do this). Lightly season and oil the heart with 1 tablespoon of olive oil, rubbing it in evenly. Flip and do the same on the other side. Heat a pan over a medium to high heat. Add 1 tablespoon of oil to the pan. Once the pan is hot, place the heart in and cook until crusted, roughly 3–4 minutes, and then flip and do the same on the other side. As the heart is a beautifully lean muscle, make sure you don't overcook – the more sanguine the better. Once cooked, add the thyme, vinegar and another spoon of olive oil. Remove from the heat and let it rest for 10 minutes whilst preparing the salad.

Wash the salad leaves and lay them round the lip of your plate. Continue to build until all the leaves have been placed, which should take about 10 minutes.

Once that is done, finely slice the ox heart, lay it over the salad and then spoon over the salsa verde. Drizzle over a little olive oil and serve.

Wild rabbit, cider & thyme stew with wild garlic polenta

The door swung wide open and in came our host, rabbits held in each hand. For a moment, it felt like we had gone back in time, for there is something ancient about this, a feeling that lasted but a second as our host went about the business of dressing the rabbit so I could go about the business of cooking it. This recipe, which suits wild rabbit, is braised in the farm's own Gilt & Flint cider, with herbs picked from around the farm, thyme being very much to the fore.

3 wild rabbits
4 celery sticks, including their leaves, trimmed
6 garlic cloves
3 carrots, trimmed
3 onions
4 tablespoons olive oil
4 bacon rashers (slices)
a small sprig of thyme, leaves picked
1 large tablespoon apple cider vinegar
1 litre (1¾ pints) rich chicken, rabbit or guinea fowl stock
1 litre (1¾ pints) dry cider
salt and pepper

POLENTA
350g (12oz) polenta
2 litres (3½ pints) water
a large handful of wild garlic
75g (2½oz) butter
175g (6oz) Parmesan cheese, finely grated

SERVES 8
Depending on the weight and age of the rabbits, the dish can be prepared in 45 minutes and braised gently for 2–3 hours

METHOD
Take the skinned rabbits and remove the kidneys, hearts and livers. Place to one side, trim and remove the shoulders, legs (including the thigh) and saddles, making sure to remove any sneaky rabbit pellets (poop). If you can't come by wild rabbits, please feel free to use farmed ones.

Coarsely chop the celery plus leaves, the garlic, carrots and onions. Put in a lidded pot with 2 tablespoons of the olive oil.

Cut the bacon into small strips and add them into the pot with the veg and a big pinch of pepper. Cook for roughly 25 minutes, or until soft.

Place the rabbit legs in a large bowl and anoint with the remaining olive oil and a large pinch of salt.

Heat a frying pan and colour the rabbit legs until rich and brown, 5 minutes per side. Sprinkle over the thyme and add the apple cider vinegar, then cook for another 2 minutes before adding the rabbit legs to the vegetable pot. Cover with the stock.

Add the cider to the now-empty rabbit pan and stir, lifting all the remaining juices. Reduce the cider by half and add to the pot. Simmer for approx. 1½–2 hours, cooking very slowly until tender – check often as wild rabbits can vary.

When the rabbit is tender and cooked, take it out and set aside. Cook down the stew for a further 10 mins over a moderate heat.

Meanwhile, in a large pot, add together the polenta and water and whisk over a gentle heat until it simmers. Leave on a low heat, stirring it regularly, for roughly 45–50 mins.

Slice the wild garlic finely, along with the stalks and flowers. Stir the wild garlic, butter and Parmesan into the polenta and add salt and pepper, to taste. Serve the polenta with the rabbit.

Pulses & grains with nettle, spinach & wild garlic

When walking around the farm, with a basket in hand for our foraged bounty, we stopped at clumps, patches and rises of leaves along the tracks, at the borders of gardens and fields and into the wild lands abutting the farm, careful not to trample lovely things underfoot. There is also a brewery on the farm, and Harry is endlessly experimenting with grains, some of great antiquity, such as the wonderful rye used in this recipe.

500g (1lb 2oz) grain, such as lentils, Barley or wild rice (I used Haye Farm's St John's Rye)
2 tablespoons water
2 tablespoons olive oil
2 handfuls of spinach
2 handfuls of tender nettle tops
a handful of wild garlic
6 free-range eggs (duck, chicken, quail)
12 toast slices
salt and pepper

SERVES 6
Cooking time 45 minutes–1 hour

METHOD
Wash the grain and cook in 2–3 litres (3½–5¼ pints) of water, bringing to the boil and letting simmer for an hour. If using St John's Rye, bring the pot off the flame and let sit for 30 minutes.

In a wide shallow pan, add the 2 tablespoons of water, 2 of olive oil and a pinch of salt and pepper.

Coarsely chop the spinach, nettle tops and garlic before placing all the leaves in the pan to simmer over a moderate heat. Cook until fully wilted – about 5–6 minutes.

In a separate pan, boil the eggs (in our case chicken eggs) for 5 minutes exactly before running under cold water.

Once the leaves have wilted fully, add them to a food processor with salt and pepper to taste. Once rendered smooth, spread the mix on a tray to cool quickly. Gently combine the rye and the nettle, spinach and wild garlic purée, ready to spread on toast and serve with a halved egg.

Braised ox cheek & Haye Farm stout

Ox cheek is such a lovely cut, with a remarkable yield that braises beautifully and, in this instance, with our host's stout, again from the farm's brewery, giving a rich savour to the dish that is eminently moreish. To give added depth and flavour, we added an oxtail to the braise, making heroic use of all the many cuts that result from a whole carcass.

4 carrots
3 small onions
6 garlic cloves
4 celery sticks
1kg (2lb 4oz) ox cheek
5 tablespoons olive oil
1 small oxtail (optional)
660ml (22fl oz) Gilt & Flint stout
4 bay leaves
a small sprig of thyme
salt

SERVES 8
Cooking time 3–3½hours

METHOD
Peel and coarsely chop the vegetables.

Trim and cut the cheeks into large pieces, approx. 3cm (1¼ inches) squared. In 1 tablespoon of the olive oil in a wide frying pan, cook the trimmings to render the fat. Once the fat is rendered, either add the trimmings to a stock pot or feed to dogs.

Lightly salt the ox cheeks and lay in the rendered fat. Cook until crusted and the ox cheek is browned all over. And if using, cut the oxtail into pieces in a similar manner then cook.

Alongside, in a heavy-based pan, warm the remaining olive oil and brown the vegetables.

Add the browned meat to the vegetables and pour 1 bottle of stout into the empty frying pan, pulling up any remaining meaty matter adhering to the bottom of the pan. With a wooden spoon, stir until the stout is reduced by half, then add to the pan of meat and vegetables with the second bottle of stout, bay leaves and thyme.

Bring the pan to a simmer and add water if required to cover. Place a tight-fitting lid atop and simmer for 2 to 2½ hours, checking it from time to time and spooning away any foam at the surface as it gathers, until tender.

Lemon, ricotta, vanilla cheesecake with malt, almonds & rhubarb cordial

We walked past the brewery and scooped up a few bowls of spent brewer's malt and this we mixed with nuts, seeds and honey, to bake crisp and shatter over the ricotta, which would have been served with rhubarb had the children not scoffed the lot!

RHUBARB
5 rhubarb sticks
2 strips of orange peel
1 vanilla pod, split and seeds scraped out
Zest and juice of 1 orange
3–4 tablespoons honey or maple syrup
300ml (½ pint) water

CHEESECAKE
100g (3½oz) blanched almonds
100g (3½oz) pine nuts
100g (3½oz) sunflower seeds
100g (3½oz) spent brewer's malt or flour or rolled oats
100g (3½oz) butter
2 tablespoons clear honey
300ml (½ pint) double (heavy) cream
750g (1lb 10oz) ricotta
finely grated zest and juice of 1 lemon
1 vanilla pod, split and scraped

SERVES 6
Cooking time 1 hour

METHOD
Preheat the oven to 150°C (300°F), Gas Mark 2.

Trim and chop the rhubarb and add to a heavy-based pan with the orange peel, vanilla pod and seeds, 3–4 tablespoons of honey, and the water. Bring to the boil and simmer until the rhubarb is fully softened. Pour into a sieve over a bowl, let drain and leave to one side. Should there be rhubarb enough left after straining, serve alongside or keep to eat with breakfast the next day.

Grind the almonds, pine nuts and sunflower seeds into a coarse crumb. Add to this either the malt or flour along with the butter and carefully work to a fine crumb. Line a baking sheet with parchment paper, evenly spread the crumble over the paper and dot with a tablespoon of honey. Bake for about 15–20 minutes, then remove from the oven and cool.

Pour the cream into a large bowl and whisk until it turns into soft peaks. Tip in the ricotta, remaining honey, lemon zest, juice and scraped vanilla seeds. Mix well, spoon into a handsome bowl, cover and refrigerate.

When ready to eat, pour over enough rhubarb cordial to cover the ricotta mix, break the crumble over the bowl and serve.

SUMMER

Life is abundant – it is summer and the sun is shining. We have come through the spring hunger gap and into a vast choice. As we move through the world, we make decision at every step of the way. In this chapter, we explore some of the subtle underlying choices that shape our beings. What food to put into our bodies, what to feed our microbiome and how to feel and recognise ourselves in a state of stress.

What we eat is important to make us healthy and strong and it also feeds our microbiome. A healthy gut and thriving microbiome help create serotonin and allow a dialogue between the gut and the brain through the vagus nerve. In this chapter, I interview three strong women. I ask the Queen of the anti-GMO movement, Dr Vandana Shiva, how do we make choices and know what to buy. I talk about my own emotional journey and how it has impacted my cognitive function via the vagus nerve with Kundalini teacher and psychotherapist Carolyn Cowan and ask science writer and microbiologist, Alanna Collen, questions about how we can shape our microbiome.

La Goccia creates summery dishes that reflect Dr Vandana Shiva's philosophy of thought and vibrant nature. Roberta Hall-McCarron recalls happy memories with her dishes, making the mind-body connection physical through taste and texture. Rachel de Thample's creations are brimming with ingredients for good gut health, as her enthusiasm for microbial life collides with delicious fruity flavours.

EARTH

In the early 2000s, I found myself hanging out in the SOAS university bar, where my sister Lara was studying anthropology. It was here that I absorbed fragments of conversation on social anthropology, learning about Margaret Mead and reading her book, *Coming of Age in Samoa*, an early study of the nature versus nurture debate. Lara's focus then turned to the anthropology of food and she was introduced to the work of Dr Vandana Shiva and her anti-GMO movement. Around this time, our dinner conversations turned to seed tracking, the monopolization of agriculture by a handful of transnational corporations, the epidemic of Indian farmer suicides and the GMO giant Monsanto.

At dinner parties, I often find the table is divided into those that are for genetic modification and those who stand strongly against it. Natural genetic engineering differs from genetic modification. Natural gene editing can happen when the root structures of two plants are combined, two plants are grafted together or when plants cross-pollinate. This can change the DNA of the plant and create a new species. Genetic modification happens in a lab using genetic engineering techniques, often taking a gene from another food or animal and implanting them into the DNA of a plant to give them more 'sought-after' properties. Bill Gates Foundation has promised to fend world hunger largely through genetic

CITIZENS

modification, but does it really yield more and what effect does it have on our ecosystems and human health? After a considerable amount of damaging press and many lawsuits (most of which were settled outside of court), Monsanto was absorbed by Bayer for $66 billion. They were genetic modification pioneers who sold an unrealistic dream of a superior food system, one with cracks so dark it was dubbed 'the most evil corporation of the world!'

When a farmer moves from traditional farming into GMO, they are no longer able to save seeds, as these seeds will be patented when their DNA sequence is modified, often to make the crop resilient to things like the herbicide Roundup or they are altered to become terminator seeds, making the second-generation seeds infertile. Each year, GMO farmers will buy seeds from corporations such as Monsanto, Bayer, Cargill and Syngenta, often accumulating debt through credit and repayment schemes. Generational knowledge is lost in a blink of an eye. Farmers who would have saved seeds in accordance with their soil and their microclimates, now rely on lab-altered seeds that promise to produce higher crop yields, but in many cases fail. According to a Guardian article back in 2014, the debt incurred by Indian farmers due to failed GMO crops saw over 270,000 farmers since 1995 take their own lives by drinking the same pesticides that would eventually destroy their topsoil. Healthy topsoil bustling with micro-organisms is vital for soil health, carbon capture and nutrient density within our food.

A conversation with

DR VANDANA SHIVA

There are many adjectives to describe Dr Vandana Shiva: activist, feminist, philosopher, environmentalist, thinker, founder, writer, guru, advocate and powerhouse. She is described as the 'Gandhi of grain', for her fight against the GMO movement and her support of ancient farming knowledge, seed saving and swapping. To say that Dr Vandana Shiva was a thorn in the side of Monsanto is an understatement; she was an antler horn thrust between their ribs. Vandana's early life included a degree in nuclear physics and travelling to the University of Western Ontario in Canada to complete her PhD in the philosophy of quantum theory. Before leaving for her studies, Vandana visited the Himalayan forests she loved and grew up in. They were vanishing and in their place she discovered a female-led environmental movement, Chipko. This was a pivotal moment, setting Vandana on the path of activism and towards setting up her own environmental movement in the form of Navdanya. Her name is now decorated with many awards relating to this work. As well as becoming an author of a great number of books, Vandana serves on the board of the International Forum on Globalization and is a member of the executive committee of the World Future Council.

AB How do we move away from the industrialization of food?

VS The first thing we have to change is to get away from these chemicals that were never needed. When I realized how wrong this was and how big the powers pushing it were, I said, 'I can't just do research on forests and rivers, there are enough people to do that work.' At that time, agriculture was the orphan of the environmental movement. No ecologist was looking at what was going on in agriculture, yet the biggest destruction was happening because of agrochemical warfare-driven agriculture. The thing we have to do is to remember once again how to grow good food. So where do we turn? We turn to nature because nature has done this throughout her evolution. She has grown food! We should look to our ancestors, our elders and our indigenous people, who are still growing food according to nature's laws, at peace with the Earth. So, this is what I've done since 1984.

AB Anna Boglione
VS Dr Vandana Shiva

AB A good job as well. What is the effect of removing inherited knowledge and replacing it with information generated by these agrochemical companies?

VS We are becoming empty minds so that advertising and propaganda can substitute knowing. For me, knowledge is a relationship. Because I've done quantum theory very consciously, I've wanted to transcend the mechanistic physics idea of the world as a machine to the world as a relationship, to the world as connection. That's what my PhD thesis was about: non-separation and quantum theory. This issue of interconnectedness has been broken by Cartesian thought, and then Cartesian thought taken into military form has turned life into our enemy, when all of life is one family. Instead of working with insects, instead of working with the biodiversity of plants, instead of working together, science has become 'kill the plants with Roundup and herbicides; kill the bees and other pollinators with pesticides,' and because we've stopped remembering the true knowledge, our empty minds just absorb and we repeat. I started the big organic network in India in the eighties when I was waking up to all of this. We were in Gandhi's ashram in Sumatra – he was the prophet of non-violence – and I suddenly found them spraying in the fields next to the ashram. So, I asked the manager, 'Why are you spraying pesticides?' And they said so innocently, 'You know, a French scientist came and he sprayed and the birds and the insects just dropped dead. We were impressed by the power.' I said, 'You pretend to be guardians, but you were impressed by the violent power to kill and you forgot the non-violent power to heal and work in harmony.'

When I started training in the early stages, I used to go village to village. One of the lessons I'd teach is that I'd take the fertilizer urea and I'd sprinkle a bit of it on an earthworm. It would start to die and I'd explain to them, 'This is what you're doing when you put chemical fertilizers on your land.' The farmers would say, 'Oh my gosh, we didn't know we were killing lives.' To open the meeting at Gandhi's ashram, as is so common in India, they had performed a sacred ceremony of fire, which was all 'om shanti, shanti'. I said, 'You say shanti (peace) when you're doing the opposite.' That whole team then became organic. So, we need to remember non-violent ways in order to practise non-violently.

AB   What do you say to people who believe that organic crops don't yield big enough harvests and take up too much land?

VS   Both of those are lies. I've now spent thirty-six years studying this field. In fact, biodiverse systems are land-saving systems. By growing a cereal with a pulse and a legume, it fixes the nitrogen. I am not applying synthetic fertilizers and I'm protecting the climate because synthetic fertilizers emit nitrous oxide, which is 300 times more damaging than carbon dioxide. I am not killing the fish in the ocean through the dead zones that are spreading by the day, and I am not killing the soil organisms, because the soil is living.

The other thing about an external input is that if I have to apply urea, I can't grow a pulse and a cereal together, because the requirement of the pulse is different from the requirement of the urea. With organic farming, they work in cooperation, but now they work in competition. I have visited so many universities where they've been trying and trying and trying to get the right mix and it just doesn't happen. Either one crop dominates or the other crop dominates because out of an external input you will have competition and then you'll have to have monocultures. Now, monocultures are land-hungry. Monocultures use more land to grow the same amount of food. In science, it's called the land equivalence ratio. The more biodiversity you have the more mixtures you have, and this is what Albert Howard found when he came to India and he wrote his book called *An Agricultural Testament*, where he said, 'I'm going to make the Indian peasant my professor,' and he

We cannot heal the food system by being the consumer.

says 'that while the West had no idea that plants could fix nitrogen, the Indian peasants knew and they were always doing mixtures.'

If there's a gift I have to give to the world, it is saying that yield is a wrong measure. Yields are all they measure, but what is the yield? The yield is what leaves the land; it is the commodity. The commodity is increasing, of course, because you're only growing one monoculture. So, what was the green revolution? Agronomist Norman Borlaug developed a dwarf wheat variety, with a higher grain yield. These dwarf varieties produced more grain and less straw. So you had a higher grain yield, but the straw is food for soil organisms and food for animals. Yield is a measure that doesn't assess the state of the soil. It definitely doesn't look at the farmer. And it doesn't look at the state of the planet or the quality of the food. I've written a book called *Who Really Feeds the World* for the Milan Expo that breaks down all of this. At Navdanya, we said we are not going to measure yield per acre, we are going to measure nutrition and health per acre. When you measure that, the more biodiverse the farm is, which can only be if it's organic, the more you have biodiversity intensification and ecological intensification, as well as higher nutritional values.

AB   It would be nice for the consumer to be able to make choices, to be able to choose something that has been grown properly against something like GMO. So how do we implement that? How do we give the consumer choices that they can see?

VS   We cannot heal the food system by being the consumer. Consumers were created out of a market world. We are Earth citizens and to be Earth citizens, there are three ways we can improve things without policing the hardworking farmer, which is all that this labelling does. The first thing you do is grow your own food, no matter how little, grow a garden. That means you shift from being a consumer to being an Earth citizen. Second, build a food community – the growth of food communities that know each other has exploded in recent years. The third thing you do is link to Fair Trade entities.

LA GOCCIA

La Goccia is our young and vibrant small plates restaurant in Covent Garden. It is named after my grandmother's interior design shop in Torino and was conjured into existence by my sister Lara. Our Italian heritage is very present here, coming out in the recipes and flavours. Having married into an ancient wine family, the food often takes influence from Lara's home in Tuscany, set within the hills of the Mazzei family vineyards. Her husband Giovanni Mazzei, charismatic and passionate for both food and wine, can at times be found in the kitchen at La Goccia alongside the brigade of chefs, or at the many wine events they host.

Working closely with my brother's Haye Farm, La Goccia follows the slow food nose-to-tail ethos. Lara's passion for healthy food, one transferred to her siblings, makes for a vibrant relationship between farm and table. It was obvious to me that I had to reach out to Vandana and pair the conversation to La Goccia. Vandana was a starting point of our school of thought, which is now embedded into our manifesto.

Beetroot salad, pickled currants & mustard dressing

This dish has the vibrancy of India, where often one vegetable takes on many different textures and flavours. Here we have shown the versatility of the beetroot, a vegetable that is beautifully high in iron and rich in colour.

BEETROOT
600g (1lb 5oz) red beetroot
600g (1lb 5oz) golden beetroot
600g (1lb 5oz) candy beetroot
50g (2oz) baby candy beetroot
110ml (3¾fl oz) extra-virgin olive oil
30g (1oz) chives, chopped
salt and pepper

MUSTARD DRESSING
75g (2½oz) wholegrain mustard
45g (1½oz) honey
juice of 1 lemon
2 teaspoons white wine vinegar
100ml (3½fl oz) extra-virgin olive oil

PICKLED CURRANTS
50ml (2fl oz) water
1 teaspoon sugar
50ml (2fl oz) cider vinegar
100g (3½oz) currants

SERVES 4–6
Cooking time 1 hour 20 minutes

METHOD
In three different pots, place the washed red, golden and candy beetroot and add enough water to cover them. Once they are boiling, reduce the heat and let them simmer for about 45 minutes until tender. Once they are cooked, drain them, peel, cut into wedges and set aside. With a mandoline or a sharp knife, thinly slice the raw baby candy beetroot.

In a bowl, whisk the mustard, honey, a pinch of salt, lemon juice and vinegar until nice and smooth. Carry on whisking and add the olive oil, little by little, making sure that you obtain a nice emulsion and the dressing is not splitting.

To make the currants, bring the water, sugar and vinegar up to a boil in a saucepan. Once it is boiling, remove from the heat and add the currants. Let them rest until cold.

In a large bowl, mix the cooked beetroot with the olive oil, salt and pepper and gently place the beetroot on a serving dish. Sprinkle over the drained pickled currants, drizzle with the mustard dressing, add the sliced raw candy beetroot and finish the dish with chopped chives.

Hummus & pinzimonio

This is a go-to in our household, as it's easy to share and a great way to utilize fibrous seasonal vegetables. Being incredibly healthy, chickpeas are commonly used in both Mediterranean and South Asian diets. Alkaline and high in calcium, they are perfect for those who prefer to avoid animal products.

HUMMUS
500g (1lb 2oz) dried chickpeas,
   soaked overnight in cold water
2 garlic cloves, chopped
200ml (7fl oz) extra-virgin olive oil,
   plus extra to serve
2 unwaxed lemons
Salt and pepper

VEGETABLES (PINZIMONIO)
10 radishes
3 courgettes (zucchini)
2 cucumbers
1 fennel bulb
2 large carrots
4 celery sticks
2 yellow (bell) peppers
2 red (bell) peppers

SERVES 4–6
Cooking time 1 hour

METHOD
Cook the chickpeas in fresh water for about 30 minutes until soft, then drain.

In a food processor, start to blend the chickpeas and chopped garlic, adding the olive oil a little at a time.

Grate the zest of the lemons and place it to one side, then squeeze the juice of the lemons and add it to the chickpeas. If the mixture is too thick and it's getting hard to blend it, you can add a bit of water, then season to taste, adding in the lemon zest.

Wash all the vegetables. Cut the radishes in half, and place them to one side, then cut the courgettes and cucumbers in 4, lengthways, and remove the seeds. Cut them again into sticks.

Chop the fennel into thin wedges, peel the carrots and cut them and the celery sticks into the same size as the courgettes and cucumbers. Cut the peppers in half, remove the seeds, and cut them into strips.

In a large bowl or plate, place the hummus and on top insert all the vegetables in a pyramid or whichever shape you desire. Drizzle with olive oil, salt and pepper and serve.

NOTE
You can also use good-quality organic tinned chickpeas; 3 tins will be enough.

Roasted aubergine, pepper, courgette agrodolce, sultana & pine nuts

The *brinjal* (aubergine) is a staple in India. Versatile and delicious, it is used in many curries. Genetically modified aubergines were developed to kill the vegetable's main pest, but following concerns raised about health and biodiversity, in 2010 Vandana led a movement that led to its ban in India. Here, we have used organic aubergine to accompany other zingy flavours.

250g (9oz) honey
500ml (18fl oz) red wine vinegar
a pinch of dried chilli
50g (1¾oz) sultanas
50g (1¾oz) pine nuts
4 aubergines (eggplants)
100ml (3½fl oz) extra-virgin olive oil,
    plus extra for the courgettes
2 red onions
5 courgettes (zucchini)
2 red (bell) peppers
10g (¾oz) marjoram
salt and pepper

SERVES 4
Cooking time 1 hour 10 minutes

METHOD
In a saucepan, bring the honey, vinegar and dried chilli to a simmer and cook until reduced to a third, making sure it doesn't caramelize. You should obtain a thick, viscous agrodolce sauce.

Soak the sultanas in lukewarm water. Toast the pine nuts in the oven at 160°C (325°F), Gas Mark 3 for 8 minutes until golden brown.

Turn up the oven to 190°C (375°F), Gas Mark 5. Cut the aubergines in half lengthways, leaving the stalks intact. Score the flesh with a knife to make a criss-cross pattern. Brush the flesh side with 2 tablespoons of olive oil and season with salt and pepper. Bake them, flesh facing up, along with the whole onions for 25 minutes. Once the onions are cooked through, let them cool then peel and cut into quarters.

Turn up the oven to 220°C (425°F), Gas Mark 7. Slice the courgettes 2cm (3/4 inch) thick and toss them with olive oil and pepper. Cut the red peppers into cubes and repeat as for the courgette. Roast the courgettes and peppers, removing the courgettes after 10–12 minutes and the peppers after 18–20 minutes.

Toss the onion, courgette and peppers in a pan over medium heat. Add a pinch of salt, sultanas and most of the sauce. Let it caramelize a bit, then add the pine nuts. Serve with the aubergine as a base, with the other vegetables on top. Sprinkle with fresh marjoram leaves and an extra drizzle of sauce.

Ricotta & lemon tortelli, butter & marjoram

Knowing where your grain comes from and having the opportunity to use organic and ancient grains in your cooking is important for your health and the biodiversity of the planet. In these tortelli, the acidity and zest of the lemon complements the creamy ricotta.

Khorasan wheat, also known as kamut flour, is an ancient grain and naturally lower in gluten. As well as boasting an array of minerals and vitamins, the grain is known to contain polyphenols. Polyphenols have strong antioxidant and anti-inflammatory properties and they help boost gut health and your overall well-being.

FILLING
1kg (2lb 4oz) sheep ricotta
200g (7oz) Parmesan cheese, grated
juice and zest of 4 unwaxed lemons
50ml (2fl oz) extra-virgin olive oil
salt and pepper

PASTA
1kg (2lb 4oz) kamut flour, plus extra for
    dusting
500g (1lb 2oz) free-range eggs

SAUCE
200g (7oz) unsalted butter
10g (¾oz) marjoram
50g (2oz) Parmesan cheese, grated

SERVES 10–12
Cooking time 2 hours

METHOD
In a large bowl, mix the ricotta, Parmesan, lemon zest and juice and olive oil, adjust with salt and pepper and place to the side.

To make the pasta, make a nest with the flour on a clean work surface, adding the eggs into the centre and using a fork to gently break up the eggs. Try to keep the flour walls intact as best as you can.

Next, use your hands to gently mix in the flour and continue working the dough to bring it together into a shaggy ball. Then, knead! At the beginning, the dough should feel pretty dry, but stick with it. It might not feel like it's going to come together, but after 8–10 minutes of kneading, it should become cohesive and smooth. If the dough still seems too dry, sprinkle your fingers with water and continue kneading to incorporate it into the dough. If the dough becomes too sticky, dust more flour onto your work surface.

When the dough comes together, shape it into a ball and wrap it in clingfilm (plastic wrap). Let the dough rest at room temperature for 30 minutes.

After the dough has rested, slice it into four pieces. Use a rolling pin or your hands to gently flatten one quarter into an oval disc and run it through the widest setting of your pasta maker. I run the dough through the pasta maker three times on this setting before proceeding to the next step.

Next, fold the dough, if you want to. This step is somewhat optional, but it will make your final pasta sheet more rectangular, which will yield longer strands of pasta. Plus, it's super simple! Just lay the dough flat and fold both short ends in to meet in the centre. Then, fold it in half lengthways to form a rectangle. Once you've folded the dough, roll it out to your desired thickness. Repeat these steps with the remaining three dough pieces. Each time you finish with a piece of dough, lay one half of it on a lightly floured baking sheet. Cut the sheet of pasta into roughly 5cm (2 inch) squares.

Half-fill a piping bag with the filling and pipe approximately 2 teaspoons of filling into the middle of each pasta square. Dip your finger in a bowl of water (or use a pastry brush if you prefer) and run it along two edges of the square and fold the square into a triangle, pressing the top together, and then work your way along

the sides. Make sure that the filling is tightly wrapped by the pasta with no air in the parcel (otherwise it may break during cooking).

Draw the bottom two corners of each triangle together to form a kerchief shape and press tightly to seal. Toss with flour, set aside on a well-floured baking sheet and cover with kitchen paper or a clean tea (dish) towel. Repeat with remaining pieces of dough.

To cook the pasta and sauce, fill a large saucepan three-quarters full of cold water, add salt and bring to the boil.

At the same time, in a saucepan melt the butter and add the marjoram. Once melted, add a ladle of the pasta cooking water and let it simmer. Add the tortelli to the water and stir a few times during cooking to prevent the pasta from sticking together. Cook for approximately 4 minutes until al dente (soft, but with a bite).

When the pasta is cooked, drain in a colander and mix it gently with the sauce. Allow this to cook for a minute. Remove from the heat, add Parmesan, then season with salt and pepper to taste.

Tomato carpaccio

This is an absolute Italian classic and representative of the simplicity of Italian cooking and how the tastes and flavours are reliant on the quality of the produce. In the spring of 1994, the first GMO product was released by Calgene to the American market: the Flavr Savr tomato. Calgene was eventually acquired by Monsanto, who later came under scrutiny for supposedly adding an antifreeze fish gene into their tomatoes. We chose this dish to pay homage to the delicious natural tomato!

1kg (2lb 4oz) Cuore del Vesuvio tomatoes
   (or any large tomato will do)
1 cucumber
2 spring onions (scallions)
150g (5½oz) red Datterini tomatoes
150g (5½oz) yellow Datterini tomatoes
1 teaspoon dried oregano
extra-virgin olive oil
sea salt and pepper
basil leaves

SERVES 4
Cooking time 15 minutes

METHOD
Cut the large tomatoes in slices and place in a single layer onto a serving dish.

Cut the cucumber in four lengthways, remove the seeds and then cut in small 5mm (3/4 inch) cubes.

Roughly chop the spring onion and cut the Datterini tomatoes in half. Place the Datterini tomatoes on top of the sliced tomatoes.

Sprinkle with the spring onion, diced cucumber and oregano.

Dress with sea salt, pepper and olive oil. Garnish the dish with basil leaves.

Pesche e vino

Another Italian classic that celebrates simplicity and fine produce, this pudding was our grandmother's favourite. She would make this for us when we were children and it would represent the coming of summer. Sharp, yet sweet flavours with the freshness of the lemon verbena picked straight from the garden.

1.5 litres (2½ pints) white wine
250g (9oz) sugar
100g (3½oz) lemon verbena
   (mint will also do), plus extra to decorate
5 yellow peaches
5 white peaches
8 flat peaches

SERVES 10
Cooking time 35 minutes

METHOD
Pour the wine into a big saucepan and add the sugar and lemon verbena.

With a knife, make a little cross on top of each peach (this will make it easier to peel once cooked).

Drop all the peaches in the wine and bring it to simmer for 5 minutes, then let it cool down.

Once the peaches are cold, remove them and cut them into pieces.

In the meantime, bring the wine back to the boil and reduce it to a syrup.

Serve the peaches in a glass and add a generous amount of syrup to cover them. Garnish with a tip of lemon verbena or some homemade sorbet.

MIND
&
BODY

Breath and movement have been shown to release trauma within the body; our accumulated history is often trapped within. Personally I am great at theorizing that same history, turning my complicated past into a linear narrative, without actually releasing the emotion. It's a wonderful skill to be able to hide from the people in front of you, and one I would say is particularly English. My following conversation with Carolyn Cowan is one of many that has taught me about the mind-body connection. Before our conversations I had little understanding about how experiencing a consistent flight or fight response, stimulated by the sympathetic nervous system, can affect one's digestive system.

My breath is tight, my vagus nerve constricted, emotions are pushed deep into my belly and out through my bones. No one ever told me to breathe. It's a familiar sentence, 'take a few deep breaths', but not one I often applied to myself. The circulation of air within the lungs, the belly expanding and contracting, the flow of chi, is our connection from out to in, from mind to body.

Carolyn's focus is on both mind and body, how we hold trauma and how, through movement and breath, we have the ability to take back control of our bodies, safety mechanisms and stress states. According to the World Health Organization, approximately 280 million people on the planet have depression and we lose around 700,000 people to suicide each year. The vast majority of those who suffer from emotional imbalances do not have access to medical help. As a species, our happiness levels are dropping – we have become more isolated, escapist and digitally obsessed. This distress can manifest itself as panic attacks, muscle pain, inflammation, brain fog, migraines and ailments that can go undiagnosed. Stemming from your brain, the vagus nerve branches out like a tree, leaves brushing against your cheeks and all the way down, sending offshoots to your heart and lungs and finally rooting in your gut and pelvic floor. It is a fascinating part of your parasympathetic nervous system, which also allows for communication between your gut and your brain. The classic term 'gut feeling' comes from our subconscious conversation between the mind and the gut. When we are in a state of unease, we have a physical response: our vagus nerve constricts, breath shortens, heart rate elevates. Research has linked childhood vagal issues due to trauma to conditions including IBS, mental illness, diabetes and autoimmune issues. In my following conversation with Carolyn, we explore what it means to let go, take back control and move away from stress states.

A conversation with

CAROLYN COWAN

I first met Carolyn a few years back. A friend of mine had put me in touch after many conversations pouring out my soul over the kitchen table. They suggested I speak to Carolyn about the mind-body connection and those conversations opened my eyes to how much trauma we hold in the body. Carolyn has a unique point of view as both a psychotherapist and Kundalini yoga teacher and she brings together the theoretical mind and the emotional body. Carolyn's work has evolved through a personal journey of addiction and trauma and an expanding career, from make-up artist to fashion designer and photographer, Kundalini teacher to psychotherapist. Carolyn founded and developed a teacher-training programme called Kundalini Global and she works as a psychosexual and relationship therapist, hosting many workshops on shame, addiction, trauma and awakening your sexual energy.

AB   Vagal issues like brain fog are debilitating. How do you start the conversation with your inner self to turn off that safety mechanism and understand what it is that you're hiding from?

CC   Brain fog comes as a defence mechanism when the charming, constructive part of the brain says, 'I can't deal with this'. You end up in the neocortex, which is the limbic aspect of the brain where the deepest aspect of the safety mechanism is stored. If you don't know how to release the hormones that cause the safety mechanism, you end up with things like brain fog, migraines, the inability to orgasm, failure to progress in labour, constant migraines, IBS and the unconscious brain reporting itself as a stomach-ache. Research has shown that during deep, restful sleep our brain is flushed with rhythmic waves of blood and cerebrospinal fluid, which clears the brains of toxins that have built up. If you don't have access to good sleep and states of being without anxiety, then you're just constantly accumulating toxins in your brain. So, then when you are hyper-stressed, they fire up this kind of brain fog. Many people, particularly at the moment, are wired to be hyper-stressed.

AB   Especially with the environment that we're in.

CC   Yes, it's also the increased screen-time everyone is experiencing, and the often violent, harmful or anxiety-inducing content that we consume. The enormous corporations behind this technology are creating products and content that encourages us to abandon ourselves, and forget who we can be. I mean, as a therapist or as a yoga teacher, it's quite frightening to watch what people are letting happen to their brain and their body. One's state of being becomes either hyper-aroused or hypo-aroused ('hyper' being very animated and 'hypo' being the opposite flat affect, a kind of not showing emotion). Both are just as dysfunctional. I remember that you and I talked about flat affect.

AB   I definitely experience the flat affect.

CC   So, it's a restricting of all the emotions. It's, 'I'm not going there and I'm not going to let you know.' It doesn't mean you don't feel them, but you're probably not yet particularly in touch with them. Over time, if somebody wants to regain their mental and physical health, they can. You've got to step over what is, in essence, an incredibly robust safety mechanism that basically says it's not safe enough to stop being freaked out.

The breath is a much faster way to take yourself back, and the breath is quite interesting because to be able to breathe well, you have to be able to control your bodily systems. So, if somebody is very stressed, either hyper or hypo, their breathing is being run by their stress system, so most people will breathe sixteen times a minute, which is fast. If you come on breath training with me, the first piece of homework is to learn to breathe once a minute.

AB   I had really intense stomach and digestive issues from the age of fifteen. The first doctor I went to was my family GP and he said, 'You're just stressed'. I then went on a fifteen-year journey to try and prove him wrong, to come all the way back round to realize I have emotional issues. I do have gut issues as well, they go hand in hand, but it's a bit like the chicken and the egg, which came

AB   Anna Boglione
CC   Carolyn Cowan

Serotonin, which is released by the relaxed vagus nerve, makes it safe enough to be still.

first? You know I can't eat gluten because it gives me a leaky gut. Did that happen because my vagus nerve was so constricted?

CC When we're anxious and stressed, when we're about to have an orgasm (if you do have one, not everybody does) if you were about to jump off a cliff dressed as a bat (you know, that skydiving thing), your whole body will be prepared. You can have very positive levels of high adrenaline and contraction, but most of us in our stress states experience a negative stress contraction. The vagus nerve is part of a whole system that gets the body ready. So, if somebody's got IBS, migraines or failure to orgasm, or they've got a need to keep peeing because they're anxious, it's all to do with the major muscle groups, the

hormones, the fascia, the amygdala. Everything works together. The vagus nerve controls all of the major organs in the body and the smooth muscle, so it goes through the throat. If I want to scream, it will be the vagus nerve that will make my eyes go up – you can feel the tension coming. If I need to suspend my breathing because I'm terrified, it's the vagus nerve that will freeze it. If I need to run or I am about to fight, it's the vagus nerve that will pull up the pelvic floor.

AB The vagus nerve is the column that everything else hangs off?

CC We can focus on the vagus nerve because the issues relative to the vagus nerve are communication, facial muscles, processing of information coming in and what you're doing with that information. If you are stressed long-term, your intestines won't work properly and you won't be secreting serotonin. So, the long-term anxiety will lead to depression. For me as a psychotherapist and for me as a yoga teacher, serotonin, which is released by the relaxed vagus nerve, makes it safe enough to be still.

AB For me personally, I feel that I have dropped into my body more so in recent years. I still feel my vagus nerve constrict, I feel the anxiety, I feel the emotion, I feel the shame. I've been trying to work with the breath, especially in finishing this book. You know, you can't go around with brain fog the whole time otherwise you can't write anything. A lot of people go around very tight, very anxious, with a shallow breath and are, in turn, depressed. We need to open up and breath.

CC I know people who don't move their bowels more than once a week. They're holding in all those toxins and there's no proper peristalsis because it's not safe enough. I also know people who've got such tight pelvic floors that they can't orgasm or they've got erectile dysfunction or premature ejaculation, but also terrible long-term constipation because they don't dare let go and they don't even know how to. So, the long-term consequences of all of it are terrible, not only on the pelvic floor and the gut, but also on the brain and brain function.

AB I've done a lot of therapy. I love more than anything to understand the way the human psyche works and I find it fascinating. Tying threads, understanding that this thing that happened to me when I was however old has triggered this and that.It's so fascinating the puzzles that we are, although no matter how much I understand it, I don't let go of the trauma that is stored in the body. I hold everything in. I won't cry in front of people, I won't let it out. I'll do everything in my power to keep constricted, my body saying, 'Nothing's going to break in. No one's going to get into this fort'. I'm trying to undo that. I'm trying to work out what one can do to release it. That is what is interesting about therapy, plus Kundalini, plus breathwork.

CC Everybody who's anxious, stressed or addicted, has trauma. They've all got the best security system on the planet running their brain. It's up to the client to decide to switch off the system or even parts of the system. It's a total oxymoron, almost incomprehensible madness, to try to take on your anxiety, because to take on your anxiety or stress, your trauma, your safety mechanisms, your brain perceives this as suicidal madness. 'Why the F would I want to show anybody my feelings, are you kidding me?' and then the therapist goes, 'Well, you know, that's going to be the path out.'

ROBERTA HALL-McCARRRON

Roberta has a unique way of bringing flavours together that subtly bounce off and complement one another, whilst alternating textures give her dishes a great depth. In the kitchen she dreams up menus, whilst her husband Shaun McCarron takes care of front of house, greeting guests and keeping the ship on course. Since opening The Little Chartroom they have won numerous awards and brilliant reviews for their restaurant. These include Best Newcomer at the Edinburgh Restaurant Awards in 2019, entries in the Michelin Guides, Breakthrough Chef of the Year at the Food and Travel Awards and entry in *The Good Food Guide* 2020; the list goes on. British *Vogue* named them best restaurant in Edinburgh, the *Guardian* says The Little Chartroom 'has sailed straight into my heart', the *Financial Times* describes it as a 'masterpiece of creative thinking,' while the *Scottish Herald* states that Roberta is a 'chef to watch'.

On meeting Roberta and Shaun, you can see that they are a relaxed yet hardworking team, with ideas brimming and a bub on the way. They are now evolving and getting ready to open their new restaurant, Eleanore. Roberta's food is light on the plate, a delight to see and to taste. I would highly recommend a trip to Edinburgh for these two.

Grilled peaches, courgette, sunflower-seed gazpacho, toasted sunflower seeds

I've taken the term 'gut feeling' and translated that, quite literally, to the feelings, thoughts and emotions that eating a beautiful dish can evoke – the sensory beauty of food and how it connects so closely with the mind. This got me thinking about those immediate, visceral reactions we experience when we eat certain ingredients, namely the nostalgia that certain dishes can evoke, be it a reminder from childhood, travel or particularly happy times. Peaches are one ingredient that has this effect on me personally. An ingredient that is rooted so intrinsically in the summertime, I can actually remember specific moments when I ate a particularly good peach, normally always outside, on hot days or trips abroad.

6 peaches
olive oil
16 courgettes (zucchini) with their flowers
fine salt

SUNFLOWER-SEED GAZPACHO
110g (3¾oz) toasted sunflower seeds,
    plus 20g (¾oz) for garnish
25g (1oz) bread, roughly chopped
¾ cucumber, peeled and roughly chopped
½ garlic clove, thinly sliced
400ml (14fl oz) oat milk
1 teaspoon olive oil
4 teaspoons sherry vinegar
1 teaspoon salt
a pinch of black pepper

SERVES 8
Marinating time 2–24 hours, cooking time 20–30 minutes

METHOD
The first thing to do is make the sunflower-seed gazpacho. Mix together the sunflower seeds with the bread, cucumber and garlic. Add the milk, olive oil, sherry vinegar and seasoning, mix well and leave to marinate for a minimum of 2 hours, and a maximum of 24 hours.

Blend the gazpacho until smooth, then check the seasoning and adjust to taste.

You can now begin to prepare everything else. Cut the peaches in half and remove the stone. Cut each half into 4 wedges, lightly salt and very lightly oil. Place on a griddle pan or barbecue for a few minutes to allow them to take on some of the char and to soften slightly.

Next, separate the flowers from the courgettes. Cut the courgettes lengthways and season lightly with fine salt and oil. Place the courgette on the griddle or barbecue and cook for approx. 2–3 minutes on each side until almost cooked.

Now that everything is ready, you can start to plate up. Place the peaches and courgettes sporadically on the plate. Drizzle generously with the sunflower-seed gazpacho. Gently tear the courgette flowers into petals and place randomly. Roughly chop the remaining sunflower seeds and sprinkle over everything.

Barbecued octopus, tomatoes, lovage emulsion

Give me an octopus and I'm immediately transported to a beach in the Med and to sharing food with friends on a warm summer's evening. There isn't another seafood that has quite the same effect.

1 x 2–3kg (4lb 8oz–6lb 8oz) octopus
300g (10½oz) lovage
400ml (14fl oz) rapeseed oil
 (not cold-pressed)
½ teaspoon sugar
3 egg yolks
4 teaspoons water
2 teaspoons Dijon mustard
2 teaspoons white wine vinegar
2 teaspoons lemon juice
1kg (2lb 4oz) heritage tomatoes
good olive oil
sea and fine salt

SERVES 8
Cooking time 2–3 hours

METHOD
Preheat the oven to 160°C (325°F), Gas Mark 3.

To begin, gently cook the octopus. Lay it in a metal or ceramic oven tray with sides. Cover with tinfoil and place in the oven for 2 hours, turning the octopus over halfway through. Check that it is cooked by spiking it with a metal skewer – there should be no resistance. If it does require more cooking, place back in the oven and check every 15 minutes.

Once cooked, allow to cool slightly. Using a knife, remove each tentacle (arm), cutting near to the head. The head can also be eaten – you just need to remove everything that's inside. Put the legs and head to one side.

To make the lovage oil, blend the lovage and 300ml (½ pint) of the oil together until it is a smooth consistency. Heat to 85°C (185°F), then immediately pass through a sieve and cool in a bowl over ice.

Next, you need to make a lovage emulsion. Mix together the sugar, egg yolks, water, mustard and white wine vinegar. Gradually whisk in 150ml (¾ pint) of lovage oil, then whisk in the remaining rapeseed oil. Add fine salt and lemon juice to taste at the end.

Now, you can start to plate up. Cut the tomatoes in various ways – some sliced, quartered and halved. Lightly dress with a drizzle of good olive oil and sprinkle with sea salt. Place somewhere warm to allow them to come up to temperature, which will allow the flavours to really come out.

Lightly oil the octopus tentacles and season with fine salt. Place on a barbecue and allow to colour and char on all sides (if you don't have a barbecue, then a griddle pan will work well too).

Lay the tomatoes on plates, place the charred octopus on top and finish with a good dollop of lovage emulsion.

Smoked pigeon breast, beetroot, orange, hazelnuts

I love to cook with the seasons and there is no season that gives me more joy than game season. I love to cook it, I love to eat it, I love to see people enjoying game for the first time. Game is low in fat and high in zinc and iron so whilst is delicious to eat, it also has health benefits too.

12 pigeon breasts
300g (10½oz) red beetroot
300g (10½oz) yellow beetroot
300g (10½oz) candy beetroot
4 oranges
3 egg yolks
2 teaspoons white wine vinegar
2 tablespoons Dijon mustard
½ teaspoon sugar
300ml (½ pint) rapeseed oil (not cold-pressed)
4 teaspoons extra-virgin olive oil
200g (7oz) salad leaves
160g (5½oz) toasted hazelnuts, chopped
salt

YOU WILL NEED
a smoking tube and wood chips
   (these can easily be found online)
a semi-deep tray
tinfoil

SERVES 8
Cooking time 2 hours

METHOD
Place the smoking tube in the tray near a window and place the wood chips into the tube, only filling one-third of the tube. Light the chips with a match and place the pigeon breasts in the tray. Make sure they are not touching the smoking tube. Cover the whole tray with tinfoil and leave to smoke for approximately 15 minutes. The tray and tube will get quite hot, so be careful.

Once smoked, remove the pigeon breasts. In a medium-hot pan, seal the pigeon breasts for approx. 2 minutes on each side. Place to one side and allow to cool.

Take the beetroots and place them separately (according to their colour) in pots and cover with cold water. Add a good pinch of salt to each pot. Bring each pot to the boil and reduce to a simmer for approximately 1–1½ hours.

Check that the beetroot is cooked with a metal skewer – there should be no resistance. Remove from the water and allow to cool slightly. Once cooked, peel the skin off (this is easier to do

when the beetroots are still slightly warm.) Cut the beetroots down to roughly 3cm (1¾ inch) pieces.

To make the orange emulsion, juice 2 of the oranges and pass the liquid through a sieve. Segment the remaining 2 oranges. Mix 2 tablespoons of the orange juice, the egg yolks, 2 teaspoons of the vinegar, the mustard and sugar together. Gradually add the rapeseed oil, whisking continuously. Season with salt at the end – taste and adjust if required.

Add the remaining white wine vinegar and the extra-virgin olive oil to the remaining orange juice. Whisk together and season to taste to make an orange dressing.

Now everything is prepared, you can start to plate the dish. First, cut each breast into 4 pieces. Place the beetroots on the plates and spoon a few small dollops of orange emulsion around the beetroots. Build up the plates with the orange segments and pigeon (each person should have 6 pieces of pigeon). Dress the salad leaves in the orange dressing and add a few leaves to each plate. Sprinkle with toasted hazelnuts to finish.

Scallop, sweetheart cabbage, crab, orange

I was very fortunate to grow up sailing on the west coast of Scotland on my family holidays. So from a young age, I was around beautiful seafood, scallops in particular. Admittedly, I did eat them deep-fried but as a child this was incredible!

3 large sweetheart cabbages, halved
2 scallops per person
    (approx. 50g/1¾oz per scallop)
olive oil
120g (4oz) butter, plus 1 tablespoon
    for the scallops
300g (10½oz) white picked crab meat
    (ask your local fishmonger for freshly
    picked white crab meat)
2 oranges, segmented and diced
30g (1oz) anchovies, drained and
    chopped into small pieces
salt and pepper

PANGRATTATO
50g (1¾oz) breadcrumbs
½ teaspoon fennel seeds
zest of 1 orange

SERVES 8
Cooking time 30 minutes

METHOD
Begin by making the pangrattato as this can be done in advance. Separately toast the breadcrumbs and fennel seeds, taking the breadcrumbs to a golden brown colour. Chop the toasted fennel seeds slightly, then mix through the breadcrumbs. Finally, zest the orange into the mix.

Blanch the cabbage halves in boiling water for approx. 2 minutes until almost cooked, then cut the halves into halves again and sear in a hot pan to get a nice char on each side.

Start cooking the scallops. Season them on both sides, then place in a hot pan with a little oil. Make sure to get a nice even, golden brown colour on each side. Finish by adding a tablespoon of butter and once it has melted, use the foaming butter to baste the scallops. This helps the scallops cook through and adds a delicious buttery nuttiness!

Gently melt the remaining butter and add the crab, orange and chopped anchovies.

To plate up, add a tablespoon of the anchovy, crab and orange mix in between the layers of cabbage, . Finish by covering the top of the cabbage with the remaining crab mix, sprinkle with pangrattato and place the scallops next to the cabbage on the plate.

Chilled apricot, ginger & almond rice pudding

Whilst Carolyn's conversation goes into some of the more negative bodily effects that stem from the mind, I believe that only joy can be experienced from eating good food, which can also be a person's most mindful moment of the day. Beautiful ingredients and joyful eating experiences are commonly an antidote to stress and sadness, and therefore food is always integral to good health, be it body or mind. Rice pudding is such a nostalgic dish for me, and for a lot of people, bringing back memories of childhood, and always decorated with a big dollop of jam. My recipe below is the more grown-up version of what has come to be one of my favourite desserts of all time; you can't but be happy when you eat it.

APRICOT JAM
375g (13oz) fresh apricots
50ml (2fl oz) water
225g (8oz) sugar
juice ½ lemon

RICE PUDDING
100g (3½oz) butter
220g (8oz) pudding rice
1 litre (1¾ pints) unsweetened almond milk
250ml (9fl oz) cream
100g (3½oz) sugar

PARKIN
100g (3½oz) golden (corn) syrup
40g (1½oz) treacle
40g (1½oz) light soft brown sugar
100g (3½oz) butter
50g (1¾oz) oatmeal
125g (4½oz) self-raising flour
½ tablespoons ground ginger
2 tablespoons milk
½ egg
10g (¾oz) stem ginger, chopped very small

ALMOND CRUNCH
60g (2oz) flaked almonds
40g (1½oz) feuilletine flakes
110g (3¾oz) almond butter
60g (2¾oz) icing (confectioner's) sugar
1 teaspoon salt
parkin

6 fresh apricots
2 packets of edible flowers

SERVES 8
Cooking time 2 hours

METHOD
All the elements to this dish can be made in advance.

To make the apricot jam, halve 375g of apricots and remove the stones. Place them in a pot with the water. Slowly bring to a gentle simmer and cook until the apricots start to break down. Add the sugar and lemon juice, bring to the boil and cook until the jam reaches the setting point – 105°C (220°F). Remove from the heat and allow to cool.

To make the rice pudding, melt the butter, add the rice and sauté for approx. 30 seconds, allowing the butter to start foaming and go brown. Add the almond milk, cream and sugar, bring to a boil and then turn the heat down to a simmer. Simmer for approx. 30 minutes, stirring regularly to avoid it catching and sticking to the bottom of the pot. Keep checking the rice to see if it's cooked. Once cooked, strain through a sieve, reserving the liquid, and place the rice in a container covered with some clingfilm (plastic wrap) to avoid it getting a skin. Do the same with the cooking liquid. Allow both to cool and then refrigerate.

Preheat the oven to 180°C (350°F), Gas Mark 4.

To make the parkin, melt the syrup, treacle, light brown sugar and butter together until the sugar has dissolved. Mix with the oatmeal, self-raising flour and ground ginger and stir into the syrup mixture, then add the milk, half an egg and the stem ginger.

Spread thinly onto a parchment paper-lined tray and bake for 10 minutes. Take out of the oven and break into smaller

pieces, then cook for an additional 5–10 minutes until crispy when cool. Once baked and cooled, chop down into smaller pieces (approx. 1cm/1/2 inch).

To make the almond crunch, toast the flaked almonds in an oven preheated to 170°C (325°F), Gas Mark 3 for approx. 10 minutes. Check and move the nuts around on the tray every few minutes. When golden brown, remove from the oven and allow to cool. Mix all the almond crunch ingredients together – you will have to use your hands to incorporate all the butter and icing sugar. Now that this is made, you can mix it together with the ginger parkin.

Gradually start adding the rice cooking liquor to the cooked rice until you are satisfied with the consistency – it should be slightly loose.

Halve the apricots, remove the stones and cut into approx. 1cm (1/2 inch) pieces.

Now everything is prepared, you can plate up. Place the apricot jam in the base of the bowl, then spoon the rice into the bowls on top of the jam. Finish by sprinkling the ginger and almond crunch on top and decorate with edible flowers.

MICROBES

On this blue-and-green planet, blue-and-green microbes called Cyanobacteria, are responsible for life as we know it. Whilst absorbing the sun's powerful rays, our planet's unique water and the carbon dioxide from our atmosphere, they create carbohydrates and in turn oxygen. Just as they were monumental in creating life, alongside mother nature, they are still vital today. Microbes have evolved with our planet, having populated it long before our modern species existed. It is said that they have been present for over 1000 times longer than we have, so it is no wonder that we have evolved closely with them.

It is only in the last decade that we have realized how our own microbiome living within the body and on its surface is imperative to our daily function. Made up of microscopic bacteria, fungi, protozoa and viruses, who can live as single-cell organisms or in clusters, these hitchhikers influence our immune system, mental health, weight, personality, autoimmune diseases and overall health. Your microbiome is shaped within the first few years of your life and is guided by your surroundings: lifestyle, nutrition, exercise, medications and relationships. We are only composed of 10% cells, the other 90% is microbes (referenced in the title of Alanna Collen's excellent book on the subject, *10% Human*) although luckily our human cells are much larger than the hundreds of thousands of species of micro-organisms living within. Our gene count is not that impressive at roughly 21,000 genes, the human genome comes below the wheat plant at 26,000 genes, the mouse at 23,000 and scrapes just above the earthworm at 20,500 genes. Could it be our microbiome that has made us such cognitive beings?. The Human Genome Project was greeted with much excitement when mapping our genes seemed to be the answer to understanding many illnesses. Now, twenty years on, mapping our microbial genomes and how they influence us and how we can influence them is causing a similar stir. Over a dodgy Zoom connection, with awful feedback, Alanna and I sat down to chat about a subject that has long fascinated me. From the altered soil structure to our immune systems, autoimmune diseases to cognitive function and even our personalities, are all influenced by what scientists thought of as 'evolutionary hangers-on'. Depending on your gut health, your microbiome may resemble a town to a metropolis, all the microbes trying to hustle and get a piece of the action. To keep your citizens happy and your streets without crime, you have to make sure there are an abundance of high-fibre, prebiotic and fermented foods.

A conversation with

DR ALANNA COLLEN

Having read Alanna's book *10% Human: How Your Body's Microbes Hold the Key to Health and Happiness*, I knew I wanted to interview her. Not knowing how, like a true millennial I took to Instagram and sent her a DM. Alanna is a BBC contributor and a science writer wo has written for the *Sunday Times Magazine* and the *Huffington Post*. She has a PhD in evolutionary biology, brims with facts and figures and is also a bat enthusiast, who can boast the discovery of bats previously unknown to science! Spending the weekend with Alanna and Rachel de Thample allowed me a glimpse into Alanna's complex brain. Listening to the ideas and research behind her upcoming book *Fatology*, whilst Rachel prepared recipes inspired by her words, was an afternoon I would very much like to repeat.

AB　We've got a rise in intolerances in autoimmune diseases and cognitive dysfunction. What has thrown our microbes out of balance?

AC　A number of things. We very much don't live in a natural way, in the sense that this is not how we evolved to live and the body is expecting the environment in which it evolved. Sensitivities to food and those kinds of things might have come about because of changes to the microbiome. In my opinion, the most significant thing to have changed it would be our diets. That's largely because we don't eat enough plants anymore. We used to have a very high-fibre diet, probably around 100–200g (3½–7oz) of fibre a day. These days in most countries, the recommended daily allowance is around 10–15g (¼–½oz), and the majority of people don't even manage that. So, we're already looking at a recommended amount that's ten times lower than what the body evolved to be used to, then when you take people's actual consumption into account, it's even lower. So, the easiest thing that anybody can do is to consume more plants for their health and for the benefit of their microbes. We also eat a lot of processed foods, which have compounds that we would never have consumed historically. I'm not talking about different food types; I'm talking about compounds that aren't technically food, such as additives and preservatives, which are altering the balance. They are not necessarily outright harmful, toxic or carcinogenic in themselves, but they alter the balance of our microbes and that changes the way our bodies function. The fibre in our diets is the first thing and the next thing is antibiotic consumption, which we know is much too high. About 50 percent of antibiotics that are prescribed in the West are unnecessary. They are crucial, life-saving drugs, but we know that we overuse them.

AB　And are antibiotics in food an issue?

AC　It's less clear how much of an issue it is, but we think so. We know it's changing the microbes of the food that we eat and we don't really know how much of it gets through to us. It's certainly an issue when it comes to antibiotic resistance, which is a related problem. There's evidence that antibiotics themselves make it into vegetable crops because the manure fertilizer that's put down often contains antibiotics from animals that have been given antibiotics as part of their regime.

AB　I imagine that the variety of fruit and vegetables that we find in the supermarkets is nothing in comparison to what we would have been eating before. We would have been eating with the seasons.

AC　I don't know if that's true. We now have access to huge ranges of fruit and vegetables from around the world in all seasons. You'd barely be able to get fruit in spring before 50, 60 or 70 years ago. We were storing apples over winter and eating jams and preserved fruit and vegetables for much of the winter and in spring before things started growing again. In some ways, we're eating more variety, but again this comes back to the concept of what we are evolutionarily supposed to eat. Perhaps we're eating things at times of the year when it's actually inappropriate for us to eat them. A truer thing to say would be that we have fewer varieties

AB　Anna Boglione
AC　Dr Alanna Collen

It is an amazing link between the microbes living in your gut and your brain function.

of vegetables that would be suitable for our locality and our season at any one time because everything is so standardized and we are consuming what the supermarkets put on offer for us, rather than the varieties we choose to sow and that our neighbours choose to sow and so on.

AB I would imagine that our bodies and our microbes would move and evolve with each season. The microbes that we need during the winter might be enhanced by preserves and ferments and the things that we have stored.

AC Yeah, that's likely to be true. We know we have circadian rhythms and we have long-acting rhythms of time that alter the way our bodies work. Some of those rhythms we are no longer exposed to, so we no longer get the urge to change the way we eat and the things that we eat. We now have access to everything all of the time.

AB Microbes also affect your personality and your energy. Can you tell me a little bit about that and maybe the experiments with mice that have gone alongside it?

AC There is an amazing link between the microbes living in your gut and your brain function, everything from your personality to your moods and mental health conditions – ranging from depression and anxiety to things that we see more as neuropsychiatric conditions, like autism, and even things like Alzheimer's and Parkinson's are affected. We're starting to understand the microbial role in those as well and much of that interaction comes via a major nerve called the vagus nerve, which runs from the brain through many organs. It collects information about what's going on in the body and feeds that information back to the brain. One of its major functions is to convey what's happening in the gut to the brain and it has, it seems, many mechanisms for doing this. A key mechanism is that our microbes release compounds from the food that we've eaten, triggering the nerve to send messages to the brain telling the brain what we've consumed and what microbes we've got. Having activity along that nerve can keep us happy or make us feel stressed, not stressed, open, fearful or vulnerable. We know that by tweaking that nerve by making it more active, in some cases of treatment-resistant depression you can effectively turn up people's happiness. We also know that eating fibre-based foods and even some probiotics can improve people's happiness levels, probably acting through that nerve as well.

Some clever scientists decided to breed germ-free mice. This means they've got absolutely no microbes living in them whatsoever, that they live in sterile bubbles, so you can colonize them with whatever microbes you want. The scientists bred two groups of mice that have different personalities, and then they swapped their microbes over. By doing this, they discovered that their personalities switched over as well. Genetically it wasn't the rodents that have these personalities, it was their microbes that brought about the personality traits. There is so much to be learnt about who we are. A lot of us like to think that who we are comes down to our own traits, our DNA, our experiences and our resilience and fortitude. Actually, it is made up of all sorts of things, including your microbes and what they do for you.

AB   I find that so fascinating because who we are is so personal to each and every individual that it's amazing to think that it's actually shared with thousands and millions of tiny little microorganisms, who contribute to our daily operation and cognitive function. I know when my gut is off my brain is really foggy and not focused. It's interesting to think that we're going to be at a point, further down the line, when maybe we can cure depression, autism and different cognitive issues through microbes.

AC   Potentially, we could help people to feel more motivated, to relax or to be happy. Also, the medical possibilities are amazing and potentially wonderful.

AB   We were talking about it earlier today how in the early 2000s they were doing the Genome Project, and now there's a similar feeling with mapping microbes.

AC   Yeah, the genome is so elegant because it is a code and it turns into a set of instructions. Microbes have a layer of complexity in that you can feed them different things, they can produce different compounds and they have multiple interactions with each other and with the body. Unravelling all of that is extremely difficult and takes many different technologies. We typically eat three to five times a day, therefore the populations are changing and the products that they're making are changing. Then there are all the interactions of the microbes and their products with ourselves and our genes. It's just mind-boggling how many connections there are and how many different variations could be going on, which is brilliant because it gives so many opportunities for addressing problems, though it is then difficult to untangle and work out why certain things happen. We've known for a long time, for example, that there's a difference between the gut microbes living in an obese person's gut and a lean person's gut, but working out exactly which microbes are doing what and why has proved quite challenging. It's not at all the fact that you can say, 'Aha, that single microbe is in everyone who's obese and everyone who's lean doesn't have it', or the other way round. It is really hard to tease apart – we've got a long way to go.

RACHEL DE THAMPLE

I first met Rachel at Petersham Nurseries when she was hosting a workshop for the chefs, teaching them all kinds of wonderful fermentation witchery. Having just started my company The Gut, I was all about microbial life and how to replenish our depleted microbiome. Rachel teaches her vast knowledge to students at River Cottage, as well as finding time to author many books. One of my favourites is *Gifts from the Modern Larder*, with perfect present-making ideas to spread the love. As well as working in the kitchens of chefs including Heston Blumenthal and Marco Pierre White, she has been a Commissioning Editor for *Waitrose Food Illustrated* and the Head of Food for the organic food box company Abel & Cole. Rachel has also served as the Course Director for the Natural Chef diploma at the College of Naturopathic Medicine.

Rachel's recipes are bursting with life, stories of her past coming through in aromatic notes. The live microbes tickle your tongue whilst enhancing your immune system and encouraging the happy good gut bacteria. Rachel and Alanna made the perfect fit, heady and hands-on, full of practice and practicalities, similarities and stories to be shared. The day we all spent together with the trading of wisdom and information could have been a whole book in itself (or at least one killer podcast).

Radish pickles with garden herbs

Radishes are so easy to grow, even in small spaces like kitchen windowsills. The beauty of them is that you get two crops in one: the root and the leaves. The pink, peppery root makes a fantastic pickle and using salt brine in place of vinegar helps their fresh, spicy flavour shine through. They are great simply fermented with a handful of garden herbs. Then, you have your second crop: the leaves, which you can also ferment after whipping them into a delicious pesto.

150g (5½oz) radish roots, tops removed
a few sprigs of garden herbs
    (thyme, bay leaves, dill)
1 garlic clove, halved
1 teaspoon sea salt
100ml (3½fl oz) water

YOU WILL NEED
200g (7oz) sterilized jar, with an airtight lid
bay leaves
grapevine leaf, horseradish leaf or a clean
    square of cloth, for weighing down

MAKES A 200G (7OZ) JAR
Fermentation time 3–21 days

METHOD
Clean the radishes. Pack them into the clean jar, tucking the herbs and garlic in as you go.

Use extra bay leaves, a grapevine leaf, a horseradish leaf or a clean square of cloth to cap the top of the radishes, which will hold them under the brine.

Whisk the salt into the water until it has fully dissolved. Pour this brine over the radishes, filling the jar right to the top (you might have excess brine).

Secure an airtight lid on the jar and set it on a plate (to hold any brine that might bubble out during fermentation). Leave to ferment at room temperature for as little as 3 days or as long as 3 weeks – the time really depends on how strong you want the flavour to be. The longer the ferment, the funkier.

Once you're happy with the flavour, transfer it to the fridge to halt the fermentation. Eat within 3 months. The pickles are lovely served with grilled mackerel or with hummus and a salad.

Kombucha blinis with honey-fermented raspberries & kombucha crème fraîche

Russian-style blinis were traditionally made with a fermented batter, yet most modern recipes have lost the overnight proving stage. However, that extra time not only makes for a richer, fuller flavour and brings the benefit of easier-to-digest grains, but the batter is much easier to work with too. I love the nutty flavour of these and, being also a bit of a health geek, I prefer them to drop scones.

200g (7oz) buckwheat flour
a pinch of sea salt
125ml (4fl oz) water
225ml (8fl oz) kombucha or jun
2 tablespoons of butter or oil, for cooking

honey-fermented rasperries (see page 130)
kombucha crème fraîche (see page 131)

MAKES 12–15
Fermenting time 8 hours
Cooking time 30 minutes

METHOD
Measure the flour into a large bowl. Stir in a good pinch of salt. Whisk in the water and kombucha until you have a smooth batter the consistency of double cream. Cover and leave to ferment overnight – the batter should be bubbly in the morning.

Heat a large frying pan over a high heat. Gloss the pan with a little butter or oil.

Drop in 1–2 tablespoonfuls of the batter per blini. Lower the heat a little and cook for about 2 minutes, until golden on the base, then flip over.

When each side has a nice bit of colour, remove from the pan and keep warm while you cook the remainder (any leftover batter can be stored in the fridge for up to 2 days).

Serve topped with kombucha crème fraîche and honey-fermented raspberries.

Honey-fermented raspberries

Raspberries and local honey are a marriage made in culinary heaven. One of my favourite moments with this preserve was when I made it with sun-kissed berries picked from the River Cottage kitchen garden on a warm August afternoon. I used honey from the River Cottage apiary, which had just been collected and fragrant rose petals from the garden. A delicious way to capture the flavours of summer.

75g (2½oz) raspberries
a handful of fragrant rose petals (optional)
75g (2½oz) honey

YOU WILL NEED
200g (7oz) sterilized jar, with an airtight lid

MAKES A 200G (7OZ) JAR
Fermentation time 7 days

METHOD
Pile your raspberries into the clean jar. Add the rose petals if using. Spoon in the honey and stir it through to fully coat the raspberries. Make sure there's a bit of headroom between the berries and the top of the jar as this ferment needs air to thrive.

Secure the jar with a lid and set it on a plate in case there is any honey overflow as it ferments, which is likely. The lid should not be completely tightened, as the fermenting process needs some air. Every day or so tighten the lid then give the jar a shake. Afterwards open the lid to let some air in, then put the lid back on loosely and return to the plate. Soon you will start to see bubbles forming.

Leave to ferment for 1 week, depending on how funky you want it to taste. As time goes on, the honey will turn a lovely pinky colour and become thinner. Pop it in the fridge once you're happy with the flavour to slow the fermentation right down. It will keep in the fridge for a few months but is best used within a month.

Kombucha crème fraîche

The most beautiful partner to everything from chocolate tart to macerated strawberries, the rich, thick tanginess of crème fraîche is just unbeatable. I love it much more than plain cream, which doesn't have the same long, lingering flavour. And it's so simple to make.

250ml (9fl oz) double (heavy) cream
4 tablespoons kombucha

YOU WILL NEED
340–500g (12oz–1lb 2oz) jar, with
    an airtight lid

MAKES 250G (9OZ)
Fermentation time 2 days

METHOD
Place the cream into the clean jar and stir in the kombucha. Cover with a clean cloth and leave to ferment at a warmish room temperature for 24 hours.

Stir, secure with a lid, then transfer to the fridge to mature for a further 24 hours. That's it! Use it straight away or store in the fridge for up to 2 weeks.

Fermented pesto

This is a fantastic example of lacto-fermentation and also a brilliant way to use up stacks of seasonal greens. One of the endless joys of this recipe is that it's completely plant-based. The lactic acid produced during the fermentation process gives the pesto the same unique umami, moreish tang that Parmesan usually imparts, meaning you can skip the cheese but still get all the flavour.

150g (5½oz) basil and/or seasonal
   herbs or greens (wild garlic, radish leaves,
   carrot tops, lemon balm, parsley, tarragon),
   roughly chopped
4 tablespoons nuts or seeds
1 teaspoon sea salt

YOU WILL NEED
200g (7oz) sterilized jar, with an airtight lid
a few extra basil leaves and/or seasonal
   herbs or greens
bay leaves
grapevine leaves or a clean square of cloth,
   for weighing down

SERVES 4
Fermentation time 3–21 days

METHOD
Blend or pound your greens and nuts/seeds in a food processor or pestle and mortar until you have a rough paste. Fold the salt in and pack it into the clean jar.

Use extra greens, bay leaves, a grapevine leaf or a clean square of cloth to cap the top of the pesto and press down firmly. The salt should draw out enough liquid to cover the pesto and the leaf cap but if not, add a pinch of salt and top up with water before sealing.

Secure an airtight lid on the jar and set on a plate (to hold any brine that might bubble out during fermentation). Leave to ferment at room temperature for as little as 3 days or as long as 3 weeks – the time really depends on how strong you want the flavour to be. The longer the ferment, the funkier.

Once you're happy with the flavour, transfer to the fridge to halt the fermentation. Eat within a month. Serve with pasta or salads, adding olive oil to finish or swirl in olive oil before using. It's also lovely blended with peas or broad beans to make a dip.

Cherry olives

My old flat in London was surrounded by cherry trees, which would ripen as the blossom on the linden trees next to them was teased out by the sun. The fragrance of both is mesmerizing and the variety of cherry near me was a really tiny, tart variety that was very much akin to olives.

I used to pickle them in vinegar or a salt brine before turning to this Japanese-inspired method, where the fruit is packed into 15g (½oz) of salt per 100g (3½oz) of fruit and then topped up with brine, which keeps the ferment airtight and safe from the growth of moulds.

200g (7oz) cherries
30g (1oz) sea salt, plus a pinch
5 black peppercorns
enough filtered water to cover

YOU WILL NEED
200g (7oz) sterilized jar or a slightly larger jar, with an airtight lid
bay leaves, linden blossom and cherry leaves or a square of cheesecloth

MAKES 200G (7OZ)
Fermentation time 2 weeks–3 months

METHOD
Wash and dry the cherries.

Mix the cherries, salt and peppercorns together. Pack into a jar. Cap with a few bay leaves, a handful of linden blossom and cherry leaves or a square cheesecloth – this provides a protective layer for the cherries to help prevent the growth of mould on top of the ferment.

Add a pinch of salt to the layer of leaves and gently trickle in enough filtered water to come right to the top of the jar.

Secure the jar with a lid, set on a plate to catch any bubbling juices caused by the fermentation process and leave to ferment for at least 2 weeks at room temperature, or up to 3 months.

You can open the jar to test the flavour after the initial 2-week period to see if you're happy enough with the flavour to eat at this point. The longer you leave them, the more the salt and fruit flavours will mingle, giving you a more olive-like flavour and texture. If you find the cherries too salty, you can rinse before serving.

The cherries are delicious in salads, tagines, paired with goat's cheese or served alongside smoked game. They also make a lovely tapenade or use them in any other dish in place of olives.

Fermented gazpacho

I love gathering gluts of over-ripe summer tomatoes from one of my favourite veg growers, Adrian Izzard of Wild Country Organics, from his stall at the end of a market day. The slightly crushed and bruised tomatoes that didn't quite appeal to the other market shoppers always take my fancy, as I know I can make this fantastic dish with them.

Fermenting tomatoes ahead of whizzing them into a gazpacho adds that extra depth of flavour, while also lending lots of probiotic goodness. When you ferment vegetables you increase their digestibility, while also making more of their nutrients available, giving this dish endless health and flavour benefits.

750g (1lb 10oz) tomatoes
1 small cucumber, peeled
2 garlic cloves
sea salt
75g (2½oz) sourdough breadcrumbs
100ml (3½fl oz) extra-virgin olive oil, plus extra for drizzling
a handful of fresh herbs, such as mint, basil, fennel and chervil, plus extra and/or edible flowers to garnish
2 tsp sherry vinegar, to taste
1 tsp raw honey, to taste (optional)
black pepper

YOU WILL NEED
1kg (2lb 4oz) sterilized jar, with an airtight lid
bay leaves, grapevine leaves or a square of cheesecloth

SERVES 4
Fermentation time 2–3 days

METHOD
Dice the tomatoes and cucumber and place in a large bowl with the garlic. Weigh the veg and calculate 4% sea salt i.e. for every 100g (31/2oz) veg, you'll need 4g (1/8oz) sea salt. Add the salt to the veg and stir it in thoroughly, ensuring it is evenly distributed.

Pack the mix into the jar, pressing down to compact it so there are no air pockets and filling the jar as close to the top as you can. Use bay leaves, a vine leaf or a square of clean cloth to cap the veg.

Seal with a lid and set on a plate to catch any bubbling juices. Leave to ferment at room temperature for 2 to 3 days, or until the brine that is produced from the veg is a bit bubbly and the veg has a lovely tang.

Once your veg are ready, spoon into a blender or food processor with your breadcrumbs and let it sit for 30 minutes to 1 hour to soften the breadcrumbs. Blend until smooth, adding the olive oil little by little as you blend until the gazpacho is smooth and looks quite creamy (due to the acid from the tomatoes and the fermentation process), emulsifying it with the oil.

Blend in a handful of herbs too (saving some for garnish). Taste and balance the flavours with a hint of sherry vinegar and/or honey and lots of black pepper, as you wish. Serve chilled, garnished with herbs, edible flowers and a finishing drizzle of olive oil.

Strawberry & elderflower kefir

Effervescent, elegant and, to my mind, as close to rosé wine as you can get without alcohol, this fruity summer water kefir is definitely my favourite virgin tipple and it's easy to make. Water kefir has been consumed for hundreds of years. Its origins trace back to the 1800s in Mexico, where the clear, jewel-like grains (bursting with gut-friendly bacteria - up to 32 different strains) were used to ferment a drink from the sweetened juice of the prickly pear cactus.

Beyond its fabulous flavour, the joy of water kefir is that it only takes a few days to brew. You can source water kefir grains online from www.happykombucha.co.uk.

6–8 freshly picked heads of elderflower
1 litre (1¾ pints) filtered water
3 tablespoons raw, organic caster sugar
2 tablespoons water kefir grains
a handful (6–7) of ripe strawberries

YOU WILL NEED
1.5 litre (2½ pint) sterilized jar, with an airtight lid
clean cloth or muslin
plastic sieve or a cheesecloth- or muslin-lined
    metal colander
1 litre (1¾ pint) sterilized bottle, for storage

MAKES 1 LITRE (1¾ PINTS)
Soaking time 24–48 hours
Fermentation time 2 days
Infusing time 1 day

METHOD
Put the freshly picked elderflower heads into the jar. Cover with the filtered water, secure with a lid and place in the fridge to infuse for 24–48 hours. Strain the elderflower heads from the water using a sieve. Discard the elderflower and whisk the sugar into the infused water until it dissolves.

Pour the infused liquor into the jar and add the kefir grains. Cover with a clean cloth or a double layer of muslin and set in a dark, cool place to ferment for 2 days. By this time, it should have developed a fizz like that of sparkling water and taste like a slightly tangier version of diluted elderflower cordial.

Strain out the kefir grains through a plastic sieve or a cheesecloth- or muslin-lined metal colander (avoid metal as it can deactivate the grains).

Pour the elderflower fizz into the bottle and discard the grains.

Quarter your strawberries (you can leave the tops on) and tuck them into the bottle with the fizz. Seal with a cork or stopper and leave at room temperature for 1 day to build carbonation and infuse the strawberry flavour into your fizz.

Strain the strawberries out or use as a garnish and serve the kefir cold. It'll keep in the fridge for up to 3 weeks, but it will become less sweet as it matures.

AUTUMN

The leaves rustle, turning red, yellow and orange, filling the horizon before falling to the ground. In the forests, fungi decompose summer's past and start to push their way up through the undergrowth. The symbiotic nature of fungi is a wonderful reminder that we all exist within spheres of connectivity to both humans and nature. When connectivity is replaced with disconnect a gradual separation occurs; overtime our landscapes become seriously altered and so do our food systems.

In this chapter Isabella Tree, conservationist and author of *Wilding*, shines a light on Britain's ever-changing biodiversity, which turns my mind to a patchwork of fields and our diminishing forests. Isabella shares tips on how we can encourage nature into our own backyards. Nutritional author Hannah Richards and I discuss how we have overlooked one key component in the inflammation crisis, while the author of *Entangled Life*, Merlin Sheldrake, gives thought to the many things we can learn from the somewhat sentient fungi, neither animal nor plant, and how they can teach us about connectivity.

Champion for nature Hugh Fearnley-Whittingstall curates recipes inspired by his friends' words, featuring primarily vegetables but with a little of Isabella's wild meat. Since I have suffered with inflammation myself, and become fascinated by the foods that help to reduce it, I have created recipes in response to Hannah's words. To round this chapter off Simon Rogan delves into the ground to concoct fungi-inspired creations.

REWILDING

My conversation with Isabella changed the way my eyes see not only our British landscape but also our neighbouring landscapes. No longer do I look out at manicured hills and arable land, just seeing the beauty of the countryside; with my newly tuned eyes, I also see what is missing. Isabella herself said that Knepp, where she lives, 'is a double-edged sword'. As a family, they moved away from farming and 'accidentally' into rewilding. Their land is steeped with fauna and flora, whilst large herbivores shape the terrain, increasing biodiversity as they move, eat and poo. They have seen a return of endangered species, such as the turtle dove, whose numbers have dropped in the UK by ninety-eight percent since the 1970s. The more common British species are also thriving at Knepp. However, living in such an oasis leaves the rest of the countryside looking pretty barren.

When I think of endangered species, my mind often jumps to the large land mammals, such as pandas and white rhinos; I don't look close to home. According to the 2019 State of Nature report, one in seven species is under threat of extinction. Have you ever wondered where the bird song has gone or the bugs that once found themselves splattered across the windscreen? Within the last fifty years, forty-four million birds have disappeared from our landscape. One-third of pollinating insects have gone too, which were vital for our existence. In the United Kingdom, most species have declined by forty percent over the last fifty years and species that are marked as a 'priority' are down on average by sixty percent – these include the skylark, Atlantic salmon, red squirrel and hazel dormouse, to name a few. The much-loved hedgehog is down by ninety-five percent, the water vole by ninety percent and the common toad by sixty-eight percent. The State of Nature report, put together by over fifty wildlife and conservation organizations in the UK, points the finger at agriculture as a key driver, saying that it is 'poisoning' the land.

We have pushed the natural world to its limits and this is becoming more apparent as we continue to use up mother nature's resources. With an increase in global trade, consumption and population seventy-five percent of the Earth's ice-free land has been significantly altered, and we have lost eighty-five percent of the Earth's wetlands, according to the 2020 WWF Living Planet report. Seeing the UK from the sky, we get glimpses of forest in amongst the tapestry of farmland. Before talking to Isabella, I'd never thought of wildlife corridors for the UK – to me, this had seemed like the perfect solution for Indian elephants or the larger endangered mammals that my mind knows so well. Creating corridors by connecting habitats such as hedgerows, woodland, meadows and wetlands together, and providing protective buffers from pollution and disturbance around them, allows species to move safely through the landscape. They can then respond to climate change and rising temperatures and reach other wildlife populations, ensuring genetic diversity and boosting population resilience. In this conversation, Isabella shares a few thoughts on how we can all be stewards for the planet, even in our back gardens.

A conversation with

ISABELLA TREE

As our connection stabilizes, I find myself zoomed into Isabella's office. It is, as you might imagine, filled with books, maps, drawings and other artefacts that represent a curious mind. Issy (as she is known) was an award-winning writer long before she won the Richard Jefferies Society Literature Award for *Wilding: The Return of Nature to a British Farm*, and is ready for my questions. We talk mainly about her journey from farmer to conservationist and rewilding Knepp, her husband's ancestral home and the inspiration for her book. Isabella's connection to nature, but more so to the land, is infectious.

AB   Anna Boglione
IT   Isabella Tree

AB   You talk about the earthworm in your book, referencing quotes from Darwin to Cleopatra. It is interesting that for such a long time people have been thinking about this little worm, what it creates for us and for our ecosystems, how it changes the course of the soil, constantly pulling leaves in, churning them up and pooing them out, so that other microbes and plants can absorb the nutrients. These vital worms, through the use of pesticides and herbicides, are being killed off. Do you think that farming with education could go back to using natural means?

IT   Yeah, I think this is what's happening right now. We're at a point, just like we were with fossil fuels and alternative energy about twenty years ago, where we've got regenerative agriculture coming down the road at us. Farmers all over the world, who are on the front line in areas of brittle zones, where their backs are against the wall, where the land is so depleted that they have to think differently and start working with nature, are beginning to move into regenerative agriculture. This is happening because farmers are realizing that if they start using regenerative methods, and particularly if they stop ploughing, they will start growing topsoil again, biodiversity will rocket and the ecosystem start to attract moisture. They get better rainfall with tree cover, and all this helps their crops and increases their yields, with low inputs – if you are not using costly chemicals, no artificial fertilizer, pesticides, herbicides and the res – then you're obviously going to be making more of a profit. The problem we have at the moment is that the big food and farming industries are so massive, so powerful, that just like the fossil fuel industries, they're trying to slow this process down. They can't stop it, but they want to put the brakes on as much as possible The change will happen, but the question is, is it going to happen sooner rather than later?

AB   Do you just regenerate topsoil by rewilding it?

IT   You ought to read Gabe Brown. He also has some very good presentations on YouTube. He explains how topsoil is regenerated using regenerative agriculture. The same thing happens under rewilding. The amazing thing about soil is it can grow and it can deepen – and much faster than most people think. We have actually just started a regenerative agriculture project at Knepp, on an area of land bordering the rewilding project. We're hoping to illustrate how regenerative agriculture and rewilding work hand-in-hand together

AB   What do you see as the ultimate vision for the UK?

IT   Well, we have to shift from conventional farming into regenerative agriculture. Everywhere that we farm, it has to be done regeneratively, whether that's no-till, cover crops and crop rotation or forestry, permaculture, silviculture and forest gardening. We know we can produce as much food using regenerative systems as current conventional agriculture. Where you've got marginal land that isn't appropriate for food production, we can rewild, so you have green corridors and margins threading through agricultural areas, and rewilding river systems and catchments. You can use rewilding as the webbing to connect these landscapes together because if we don't, species will be isolated and unable to move in the face of climate change. They'll also be genetically segregated from wider populations. Eventually, species in isolated places are doomed to fail if they cannot connect with other areas of nature. Having areas of nature around crops also boosts the yield because these nature margins provide pollinating insects, natural pest control, micro-climates, flood mitigation and safe water storage – you're replenishing your water tables and releasing pure water, clean water, onto agricultural land in a safe way. Your rewilded areas can also act as buffers against extreme weather events, which we're going to get more and more of with climate change. Rewilding really works hand in hand with regenerative agriculture. It's a way of boosting our yields, but also protecting our food production areas, as well as increasing biodiversity and human access to nature.

AB   Did you find in your process of rewilding that you had your own inner rewilding journey?

IT   Yes, definitely. It completely changes your psychology. We were control-freaks, micromanaging, super-tidy farmers, and now we've become a bit more relaxed, certainly trusting that nature can do its thing without us. It's a completely different mindset, being able to live with the unpredictability of nature, the boom-and-bust scenarios, and not panic when you suddenly see ragwort and creeping thistle taking over dozens of acres. You know that nature is going to make its own balance. So, it's actually a very Zen way of living. If you can live with rewilding, you accept change as a very positive force and it's no longer frightening. It's an adventure and something hugely creative.

AB   Were there moments where you felt panicked?

IT   Yes. I describe it in the book, when we had that first outbreak of creeping thistle, that was a panic, but just by allowing it to be, we had this miraculous apparition of tens of thousands of painted lady butterflies that arrived and laid their eggs on the creeping thistle and eventually eradicated every plant. What is a boom year for one thing is going to become a boom year for something else that usually redresses the balance. The more dynamic a system can be, the more you can let go and just allow things to happen, and the more exciting it is for biodiversity.

AB   It's also interesting that as humans we try to manipulate nature in every element, like repopulating rivers with fish from fisheries. We think we've got the next thing to fix something. With time, you realize that by changing one thing, subsequently you change a thousand others – everything is connected.

IT   I think it's a lot to do with the mechanical mindset and the lack of observation of nature as we've become so disconnected from it. We have a lot to learn from indigenous peoples. There are hundreds of different cultures out there in the world that have been living sustainably for thousands of years. In Papua New Guinea, they have a very sophisticated system of horticulture involving mounds of compost, protecting plants from frost, and using pigs for tillage and to fertilize the soil. I recently guest-edited an edition of *Granta* called Second Nature; one of the essays was by one of these leading regenerative agricultural farmers in Australia called Charlie Massy. He describes his epiphany and how he realized that what he was doing was so destructive. He has now completely changed the way he farms. One of the contributing factors to his epiphany was conversations with an Aboriginal leader called Rod Mason, who also wrote an essay for this issue. Rod's understanding of how ecosystems work and function in Australia is invaluable. He understands how on that brittle, thin soil of Australia, most life depended on a layer of fungi that covered the soil and held it together, attracting the moisture for all these other species, and now that fungal layer is gone. It may be possible to get it back. The knowledge is still out there – we just have to listen.

AB   And slow down ever so slightly. I really liked you mentioning that Sussex folk had thirty-five words for mud. It shows that we had the knowledge as well and we were once so interconnected with the land that we too have words to describe our landscapes. Due to the pace of life, we have become ever so disconnected. In Japan, they prescribe 'forest bathing', or Shinrin-Yoku, not only as a form of meditation but also as a medical treatment for stress and anxiety. Trees also release phytoncide oils that help increase natural killer cells, which is great as they help ward off viruses and fight cancer. It would be nice if we could redefine what we see as nature. Rolling hills are lovely but they don't give you the same feeling as being surrounded by biodiversity.

IT   We have to remember that we've been urban beings for just a tiny blip of our evolution. It's very deep-rooted in us, this connection with nature. When we sever ourselves from it, it's not surprising that we start to suffer physically, mentally and in all sorts of other ways. Covid has made that very clear. The instinctive need that people are feeling to get out in nature right now – it's an incredibly powerful message.

AB As your land became wilder, did you feel this connection deepen?

IT It's a double-edged sword in many ways because we are literally in a bubble here at Knepp – we can walk out of our front door and hear wonderful birdsong and see extraordinary things every day. I mean, there's a great white egret out there right now on the lake. It's a species that is just arriving in Britain. It's a pioneer and there are not many places you can see that. We've got white storks wheeling overhead. There's always something extraordinary happening. You go for a walk and you just feel fantastic because you're surrounded by life doing its thing. Then you go for a walk somewhere else, like the South Downs where you used to love walking before, and you notice what isn't there. The lack of insects, the sounds that aren't there, the birds that should be there but aren't. You recognize how in much of the world this isn't happening – yet.

AB What can the individual do to help biodiversity and draw nature into their own little patch?

IT Charlie and I are writing *The Book of Rewilding*, which is going to be a practical guide on how to rewild big and small and how everybody can do their bit. One thing is obvious – stop using chemicals of any description, and then amazing things will start to happen. If you start thinking, 'How can I do this sustainably', it's amazing how little you need to go and shop for. You can exchange plants, there are seed Sundays where people are starting to swap seeds and you can exchange cuttings and homegrown plants that have not been propagated using pesticides, as they usually are in commercial nurseries. So, there are lots of green things you can do. One of the most important things about the rewilding ethos is connecting. For example, I have a friend in Bristol who was a film cameraman and knows his nature very well, so he thought he'd rewild his back garden. He found himself with twenty-five grass snakes – that's the kind of habitat he was able to create. But what he did, which is much more special, was he talked to all his neighbours in that street. So, they all understood what creating a habitat for wildlife meant. One of them had a pond, another a beetle bank, another had some sandy habitat, another had low-lying herbs, and someone else had some thorny scrub. Then he persuaded them all to cut holes in their hedges and fences so that hedgehogs and small mammals could move between them. Suddenly, that row of gardens becomes a wildlife corridor and if you can connect that row of back gardens with an inner-city park at one end and maybe an embankment, railway or canal at the other end, you've got a real flow of movement of wildlife coming into cities and out again. We've got to think about how we connect, so that we're not just an isolated patch of land or back garden, vegetable patch, farm or whatever. It's always thinking, 'How can I generate more dynamism from this by connecting with other nearby areas?'

AB That's true. You don't think about the roads, motorways and fences blocking wildlife. How on earth is a human going to navigate such blockages, let alone an animal? For people who might not have that much space, but want to grow something to attract insects and bees, what would you recommend?

IT Well, I think you have to be hugely careful. It's absolutely iniquitous that you can go to the garden centre and buy a plant that has a tag on it saying 'bee-friendly' that has been propagated using pesticides. So, when you plant it in your garden, the pollinator comes along very happily doing its job and is killed. You're attracting pollinators and killing them! We have to be really careful about where we get our seeds from and where we get our plants from. They have to be organic, preferably by swapping, so you are swapping knowledge as well as connecting with your neighbours.

HUGH FEARNLEY-WHITTINGSTALL

Driving to a retreat site visit in Cornwall, an email flies into by inbox from Satish Kumar, introducing me to Hugh. A speedy response ensures that on my way back I can pop in to say hi. Hugh is warm, his character kind and approachable. He likes the project and knows Isabella, so it is a fit. Growing up we all knew Hugh from the telly, fighting for fish, the environment and good healthy food. More recently, Hugh's documentary crusades include *Hugh's War on Waste* and *Britain's Fat Fight*. His TV series *River Cottage* captivated us, taught us and raised questions on how to be sustainable. Nowadays, River Cottage plays host to wonderful chefs, extraordinary cookbooks and awe-inspiring workshops.

 Arriving again at his smallholding farm, this time to shoot his recipes, the dogs greet me like a long-lost friend, excited and enthusiastically wanting affection. Hugh appears, apologizing for his motley pack. Hugh's recipes, inspired by Isabella and her family's work down at Knepp, keep true to his pledge to eat more veg and only good-quality (and in this case wild) meat.

Ceps on tomato toast

Mushrooms on toast are great, and so is a classic bruschetta piled with ripe toms and good olive oil. A collision of the two happens to be very delicious indeed, and if the mushrooms are ceps, it's sensational.

a knob of butter
a small cep or two (or half a big one) or 2 large portobello mushrooms
2 garlic cloves, one sliced and one halved
1 large slice of robust wholegrain bread (ideally sourdough)
extra-virgin olive oil
a ripe slicing/beefsteak tomato (or a few cherry toms)
a squeeze of lemon juice
grated or slivered Parmesan or similar hard Cheese (optional)
salt and pepper

SERVES 1
Cooking time 20 minutes

METHOD
Heat the butter in a medium frying pan over a medium heat. Slice the cep(s) or portobello mushrooms about 3mm (⅛ inch) thick (including the stalk if it's in good nick) and add to the pan with a few slivers of garlic.

Fry for 3–4 minutes, turning occasionally, seasoning as you go, until the cep is lightly browned on both sides (portobello mushrooms may take longer – you want them tender and just a little browned). Take off the heat and keep warm in the pan.

Meanwhile, toast/grill the bread. It's good grilled on a griddle, with charred stripes, but a toaster is okay if that's what you have.

Rub the halved garlic cloves lightly over the hot toast. Trickle over the extra-virgin olive oil and season with a little salt. Slice the tomato thinly and arrange to completely cover the toast (ideally without overlapping), season and give it another little trickle of oil. Cover the tomato with slices of the still warm mushrooms.

Add a squeeze of lemon juice and a little more salt and pepper and finish, if you like, with some grated or slivered hard cheese. Eat straightaway.

Roast squash & pears with chillies & nasturtiums

This is a lovely autumn dish that captures the sweet mellowness of squash and pears and gives it a lift with the punchy heat of chillies and nasturtiums. A sprinkle of peppery rocket leaves works well instead of the nasturtiums but, of course, it's lovely to have the flowers. I've used homegrown chillies, a variety called Fireflame, which is not too hot and roasts beautifully. This makes a lovely vegan starter or can be part of a big mezze veggie spread.

1 medium butternut, onion or acorn squash, or ½ a large one, about 800g–1kg (2lb)
2 small onions or 3–4 shallots, peeled and quartered
3–4 tablespoons rapeseed or olive oil
2 pears (or apples), sliced into quarters or sixths and cored
up to 8 large chillies (if mild, cut lengthways and remove seeds, if hotter, use fewer and chop smaller)
a few sprigs of rosemary (optional)
6–8 garlic cloves, skin on, lightly squashed
a handful of nasturtium or rocket leaves, plus flowers if you have them
a dash of extra-virgin olive oil
a squeeze of lemon juice
sea salt and pepper

SERVES 4
Cooking time 1 hour

METHOD
Preheat the oven to 190°C (375°F), Gas Mark 5.

Slice the squash into wedges, deseeding it as you go (you can dry the seeds in an airing cupboard and use them to grow your own squashes). I leave the skin on most squashes when I'm roasting them, but you can peel it off if you prefer. Put them in a small roasting tin with the shallots/onion, season well with salt and pepper, trickle over 2 tablespoons of the oil and toss the lot together.

Place in the oven for 20 minutes, then remove, add the pear, chillies, rosemary (if using) and garlic, toss again with 1 more tablespoon of oil, and return to the oven for another 20 mins or so, till all the veg is tender and the onions and squash are nicely caramelized at the edges.

Remove from the oven and leave to cool to room temperature. Lightly dress the leaves and flowers with a dash of oil, a squeeze of lemon and a little salt and pepper and arrange over the dish. Finally, scatter over the flowers.

Ale-braised short ribs

Beef is not an everyday food and this is not an everyday dish. However pastured cattle can, and should be, part of a regenerative approach to agriculture, and are also key to restoring habitat in rewilding projects, as Isabella Tree has argued powerfully in her game-changing book, *Wilding*. And so, it was a pleasure to put together this lovely slow-braised dish, with amazing short-rib from the Knepp herd of Longhorn cattle.

2 tablespoons vegetable oil or beef suet
1.2kg (2lb 10oz) trimmed short rib of beef, in 6–8 large pieces, on the bone
1 litre (1¾ pints) good brown ale or strong, well-hopped beer
2 bay leaves
3 large carrots, peeled and chunkily chopped
3 onions, peeled and quartered
4 celery sticks

You can also make this dish with red wine instead of beer – 500ml (18fl oz) of wine made up to 1 litre (1¾ pints) with water or veg stock.

SERVES 6
Cooking time 3 hours

METHOD
Preheat the oven to 140°C (275°F), Gas Mark 1.

Heat half the oil or beef fat in a large frying pan over a medium heat. Season the chunks of short rib with salt and pepper, then transfer them to the hot pan and brown them well all over for at least 5 minutes – you want some good colour on them, as this is where the flavour comes from. If you don't have a really big frying pan, you may have to do this in two batches. Transfer the browned meat to a wide casserole or fairly deep ovenproof dish.

When the meat is browned, pour a glass of the beer into the still-hot pan to deglaze the burnt bits (by scraping with a wooden spatula) and heat up the beer. Add all the beer and bring to a simmer, then pour the hot beer over the beef. It should more than half cover the beef. You might not need it all straightaway, so keep some for topping up the braise. Tuck a couple of bay leaves into the dish and put in the oven.

After about 40–50 minutes, remove from the oven and turn over the pieces of short rib, then return to the oven for another 40–50 mins.

Now, prep the veg. You're adding it later so that it's not completely mushy when the beef is totally tender. There's still plenty of time for it to flavour the liquor of your braise. Heat a little more oil or beef fat in a pan and brown the carrots and onions, as you did the meat. Don't be shy of putting some good colour on the veg, even burning them a little – it's all good flavour.

Remove the meat from the oven and turn the pieces one more time, this time adding the browned carrots and onions and the celery (no need to brown this) and topping up with a little more beer or water if you think it needs it. Check the taste of the liquor and add a touch more salt and pepper, if you like.

Return to the oven for another hour or so, until the meat is completely tender. Serve each person on warmed plates or bowls, with one large or two smaller pieces of short rib, and plenty of the liquor and veg. It's great with creamy mashed potatoes and some simply steamed greens. It's fine (arguably better) to cook this dish a day (or two, or three) ahead, and let it go cold. Keep covered in the fridge.

You can lift off some of the hard set fat from the top, if you like, before reheating, then gently reheat at 140°C (275°F), Gas Mark 1 for at least 40 minutes, until completely hot.

VARIATION
ALE-BRAISED BRISKET

To feed a crowd, this dish also works brilliantly with a single large piece of brisket, anything up to 2.5kg (5lb 8oz) (or whatever fits in your largest ovenproof dish!). Scale-up quantities of the veg and beer a bit and brown the beef really well in your biggest pan or roast it for 20–30 minutes at 210°C (410°F), Gas Mark 6 to get some good colour on it before braising. Then proceed as above, turning the meat every 50 mins or so.

When you've browned the veg halfway through the braising of the brisket, just tuck it all under the beef. Depending on the size of your brisket, this may well take longer than the short-rib version – 3½–4 hours should do it.

Cabbage & bean soup

I love the parsimony of this soup, but cabbage and beans (bolstered by the classic stock veg – onions and celery) have easily enough going for them to star in a soup. I particularly like to make it with 'demi-sec' autumn beans from the garden: overgrown French and/or borlotti beans that have gone past their whole-pod, green-bean moment, but can still be harvested for the creamy beans within. This also works well with tinned beans, such as butter beans, cannellini, borlotti or a mix.

1 large or 2 medium onions
2–3 celery sticks
2–3 garlic cloves
2 tablespoons rapeseed or olive oil
1 litre (1¾ pints) vegetable stock
400g (14oz) podded fresh demi-sec beans
OR 2 x 400g (14oz) tins borlotti or butter beans
½ a large Savoy (or other green-type) cabbage or
    a whole small one, shredded, but not too fine
sea salt and pepper

TO SERVE
4 thick slices of sourdough or good rustic bread
1 garlic clove, halved
extra-virgin olive oil

SERVES 4
Cooking time 45 minutes

METHOD
Slice the celery thickly, the onions thinly, and the garlic thinnest of all.

Heat a medium saucepan over a medium heat and add the oil. Add the onion, garlic and celery and sweat gently for 10–12 minutes, until soft but not coloured. Add the stock and the fresh beans, if using, and bring to a gentle simmer. Cook for another 10–15 minutes or so until the beans are tender, then add the cabbage and simmer for another 6–8 minutes till the cabbage is cooked.

If using tinned beans, you can add the stock, cabbage and the (drained, rinsed) beans to the sweated onion/celery/garlic and simmer for 6–8 minutes.

While the cabbage is cooking, toast or grill the bread, then rub the surface of the hot toast lightly with the clove of garlic. Trickle each slice generously with oil.

Check and adjust the seasoning of the soup, then ladle into warmed bowls and serve with the garlicky bread on the side or dunked in the bowl, and another generous trickle of extra-virgin olive oil over the soup itself.

VARIATION
You can build this soup up with more seasonal additions, including a handful of chopped fresh tomatoes, added near the end, and more 'carby' veg, such as a wedge of squash, diced, or roots, such as carrots, parsnips and celeriac, added with the stock and before the cabbage.

Bramley & quince fumble

In River Cottage world, a fumble is a cross between a fool and a crumble. It's a pretty good place to be. You can make the fruit and crumble elements separately, then bring them together, with a dollop of cream or yoghurt or, in this case, half and half. Any tart fruit purée works well, and you can also use crushed summer fruits, such as strawberries or raspberries, for the fruity element. Bramley apples are classic and make my very favourite autumn version of this pud. A hint of quince adds an aromatic note, but it's not vital.

CRUMBLE
75g (2½oz) butter, diced
100g (3½oz) fine wholemeal flour
50g (1¾oz) rolled (porridge) oats
    or fine oatmeal
50g (1¾oz) ground almonds
30g sugar (soft brown, golden granulated
    or raw/demerara)

BRAMLEY COMPOTE
juice of 1 lemon
1kg (2lb 4oz) Bramley or other
    cooking apples
1 small quince (optional) (or add a
    star anise instead)
50g (1¾oz) caster (superfine)
    sugar, more if needed

TO SERVE
150ml (5½fl oz) double (heavy) or whipping cream
150g (5½oz) whole natural yoghurt

SERVES 6
Cooking time 1 hour

METHOD
For the crumble, preheat the oven to 190°C (375°F), Gas Mark 5 and have ready a large baking tray.

Rub the cold butter into the flour in a large bowl to get a coarse breadcrumb texture, then stir in the other ingredients. Squeeze some of the mix into chunky clumps of varying sizes and spread all the crumble out on the baking tray. Bake for 15–20 minutes, stirring the mix at least once, until golden brown. Leave to cool completely. If you are not using it straightaway, store the crumble in a jar or tin.

To make the compote, put the lemon juice into a large pan. Peel and core the apples and slice thinly into the pan, tossing them with the juice as you go so they don't brown. Peel and grate the quince (if using) into the same pan, leaving behind the seedy core. Add the sugar and 2 tablespoons of water.

Bring to a simmer, stirring often, and then cook gently, stirring occasionally to help the apples break down, for about 20 minutes until you have a slightly chunky purée. Taste and add more sugar, if you like, but keep the compote nicely tart because it will be paired with the sweet crumble. You can either serve your compote straight away or let it cool and then chill it.

Whip the cream until fairly thick and fold in the yoghurt. Divide the apple compote into serving glasses or small bowls, add a generous dollop of the yoghurt/whipped cream combo and top with a layer of crumble mix. Tuck in!

INFLAMMATION

In my mind, there are two categories of inflammation – the first health and recovery, the second chronic and illness. Our bodies rely on their inflammatory response to protect us from pathogens, viruses and bacteria. It allows the immune system to jump into action and heal the infected area and without it, wounds would fester. John Hunter, a Scottish surgeon, wrote a book on inflammation and gunshot wounds that was published in 1794, linking inflammation to diseases such as smallpox, venereal infections and tuberculosis. Two hundred years later, the list of illnesses that are influenced by inflammation is extensive. This list includes obesity, heart disease, diabetes, inflammatory bowel disease, Alzheimer's, rheumatoid arthritis and cancer.

An increase in inflammation in the body can also cause dysbiosis, an imbalance in the gut's microbiota. A study showed that it only takes two weeks of diet change to have an effect on your gut's health! The trial asked African Americans who ate a western diet, high in protein and fats, to swap diets with South Africans who ate a diet high in fibre and low in proteins and fat. 'After two weeks on the African diet, the American group had significantly less inflammation in the colon and reduced biomarkers of cancer risk. In the African group, measurements indicating cancer risk dramatically increased after two weeks on the western diet.'

There are many studies showing the positive effects of a high-fibre diet on chronic inflammation, while other foods that help heal the gut and reduce inflammation are ferments. High-fibre foods, such as oats, apples, barley, bananas and lentils act as a prebiotic, whereas ferments, such as kimchi, miso, sauerkraut, kefir and kombucha are all probiotics, brimming with good bacteria. Fermented foods seem to have had a renaissance, even though the positive impacts have been documented since the early 1900s, when Elie Metchnikoff, a Russian-born biologist, published a paper linking longevity with fermented foods containing the probiotic lactobacillus.

Another factor that impacts inflammation is stress. Often our bodies throw forth many different ailments, which can be misdiagnosed without taking stress into account. I recently came across a book called *Earthing*, which says that walking on the earth's surface barefoot, also known as grounding, stabilizes the body to its natural electric charge. Grounding has been shown to reduce inflammation as well as improve many other imbalances. In the following interview with nutritionist Hannah Richards, we discuss a cure for inflammation that, like earthing, is simple yet powerful and a totally new concept for my mind.

A conversation with

HANNAH RICHARDS

Hannah joined me on the very first day retreat I hosted with The Gut at Petersham Nurseries. Her unique mindset when it comes to healing the body resonated, whilst encouraging my own learning. Hannah's philosophy is that 'You can't heal in the same environment you got sick in.' This guides her understanding of her patients and informs their path to wellness. Much of this can be found in her book *The Best Possible You: A Unique Nutritional Guide to Healing Your Body*.

Hannah has trained with many influential practitioners in health and well-being and is also a certified Metabolic Typing Advisor, a Functional Diagnostic Nutritional Coach and a certified and registered nutritionist from the College of Naturopathic Medicine. Hannah was previously a regular nutritionist on *The Chrissy B Show* on Sky TV, where she also cooked delicious and easy-to-follow recipes.

AB   Anna Boglione

HR   Hannah Richards

AB   Good inflammation is when, for example, you cut yourself and the inflammation allows parts of the body to heal, but chronic inflammation can cause you to have arthritis, heart disease and IBS. What are the differences between the two?

HR   Inflammation is a protective mechanism. You sprain your ankle, it swells up, you cut yourself, it swells up. Chronic inflammation is more about ischemic tissue damage and blood flow. When chronic inflammation sets in, it starts affecting every other part of the body — it's all about blood flow. By the time people get gut issues, you can't fix them by looking at the gut. You've got to understand that chronic inflammation has taken hold and there are problems with the entire body — the visceral organs. It basically comes down to posture, breathing and the diaphragm.

AB   I know that engaging the diaphragm when breathing can create a gentle massaging feeling for other internal organs, and how this affects gut health, but I hadn't thought about it in terms of being hunched over with bad posture.

HR   As an example, I often get a client that will come and say, 'I've seen seven naturopathic nutritionists and, to be quite honest, none of them has helped me.' So, as a practitioner, you have to understand the psychological place that a client is in. Have they seen seven people and they haven't helped them because the practitioners haven't had the skills to take them down a different line?

You can't fix the gut with food. By the time the gut's in a place of discomfort, it's about the digestive system and then it's about the tissues and the blood flow. This takes you back to the posture, which takes you back to the organs, and that takes you back to breathing. So, when food stops working: when supplements stop working, when testing never worked in the first place, you've got to understand that things like neuromuscular therapy work. The organs aren't flat-packed like an IKEA wardrobe — they're three dimensional and live off ligaments. If your posture is crushed, you're automatically bringing forward your rib cage and squashing the diaphragm and all the organs that are housed from the sternum to the pubic bone. So, you've got all these organs neatly, but three-dimensionally, hanging out because your posture has created this kyphosis (exaggerated rounding of the upper back) and there isn't enough movement around the organs to work. For example, the stomach will fold over the liver, which can stop the pancreas from producing enough enzymes as it is being blocked by another organ. This is all because the head comes forward, which can weigh a stone, you have a dowager's hump on the back compressing the diaphragm, and then you have a reduced blood flow around the body to the visceral organs. That's what continues the inflammation and also prevents the organs working well enough with one another, ultimately depriving the gut of enough blood flow for action and motility. The inflammation in the body stems from bad posture, but no one's looking for bad posture. They're looking for the right supplement.

AB There's a lot of research about stress causing inflammation. We are tightening up from stress, constricting our diaphragm and vagus nerve. How does this affect the inflammation within the body?

HR Imagine your physiology when angry and stressed. Your muscles tighten, you're not breathing, you're breathing high or you've got shortness of breath. Think about what that does to every other system in the body. So, when you're in that state of stress, the only thing your body needs to do is keep you alive. It's not interested in digestion. It doesn't care if you're bloated. It certainly doesn't care if you're constipated or gassy. It's just interested in keeping you alive because every bit of energy is going into those muscles. If we treat using alternative medicine the same way the allopathic model does, then we're basically saying sauerkraut for gut health, vitamin E for skincare and go to the gym for back pain. Whilst they're nice alternatives, we're not linking the dots well enough. This is why getting yourself in the hands of people who understand the body three-dimensionally, how it works and how it's linked, is vital to good health.

Breath and posture are two of the most important things when it comes to repairing the body.

AB Looking at the body as a whole, what would your advice be to someone who is suffering with chronic inflammation?

HR I'd start by looking at breathing. A really great, easy trick is to put your wallet or a teddy bear on top of your tummy, on your belly button, while lying down flat. You want to be able to see that it's moving up and down, so that you have control over the abdominal wall. The breath basically feeds every organ and a lack of understanding of the breath creates dysfunction. Yoga is one of the best things a person can do in life to understand the breath and breath and posture are two of the most important things when it comes to repairing or reducing anything in the body. If you don't get the structure or the foundations right, then none of it will work because you're refusing to look at the physiology of the body, which has been created through stressing about diet, relationships, bad eating and too much exercise, you name it.

AB A lot of childhood and adult trauma leads into vagal issues, which turn into breath and gut issues. It's fascinating the way that these simple systems are so interlinked and that we somewhat overlook them. What role does food play, because it does to a certain degree play a role?

HR By the time you're eating seasonally, organically and off the land, then it often doesn't matter what you eat, how you eat, when you eat. The hardest thing is to get your body physically feeling good, because we are compressed mentally, emotionally and physically. The emotional stuff is hard and the physical stuff is hard. People only want to heal themselves with food and it's just not enough.

One of the biggest questions I ask people when I work with them is to list the top three loves in their life — the three most important things. Very rarely is number one themselves, and there lies the problem of a person's healing process. If we can't love ourselves, we don't put ourselves first, then we won't be able to heal because we're putting everyone else or every other thing in front of us. That's what nutrition is. People laugh when I say it, but it's got very little to do with food initially. When you feel emotionally strong, mentally strong and physically strong, you don't eat rubbish, you don't overeat, you don't undereat. You feed yourself and you nourish yourself.

That's why giving someone a diet sheet can be the worst thing you do: 'Here's a plan for failure. You've already failed, which is why you've come to see me.' Instead, it's about getting a person to understand where they are, whether that's in a chaotic state or a state of bliss, and to recognize that if they are in chaos, that's not where they want to be. It doesn't matter that they're not where they want to be, but they have to be aware that where they are isn't working and then you've got lift-off. If you try and fix their diet on day one, you'll not only put a person in a worse place, but you'll extend the healing process. That's why it's about physiology and metabolism and not about whether you've got thirty different plant foods in your diet.

ANNA BOGLIONE

Having grown up at Petersham and in the Nurseries, our restaurant, I learnt how to pair tastes and texture with seasonal produce. Endless menu tastings and critiquing what works and what did not, very much influenced my own style of cooking. My younger sister Ruby and I would telepathically dream up meals together in our own kitchen across the garden.

Food became a point of curiosity, not only the way one creates a meal but also how it effect the body, it's nutrient density and health benefits. My wellness platform, THE GUT, allows me to spend time creating topic derived menus. My aim for the future is to have a fixed location for THE GUT, where I can dream up menus and be excited by healthy, beautiful, fresh, organic produce.

Chia seed pot with turmeric, ginger & stewed apple

This is a great recipe for breakfast or to finish off a light meal. Chia seeds are packed with all nine essential amino acids, healthy fats, protein and are high in both soluble and insoluble fibre, making them a prebiotic. Prebiotics are essentially food for your gut microbes and happy microbes equal happy gut.

Chia seeds, when soaked and consumed, have this gel-like quality, which helps soothe and restore your GI tract, as well as promoting your bowels' regularity. Turmeric contains curcumin, a substance that is scientifically proven to have powerful anti-inflammatory properties. Make sure you also add black pepper, which contains piperine, so your body is able to fully absorb the curcumin.

I have written this recipe per serving as it is easy and often nice to make and serve in a glass – this way you can decide how many glasses to make.

2 tablespoons chia seeds
a pinch of peeled and grated fresh turmeric
2 pinches of peeled and grated fresh root ginger
2 teaspoons honey
a pinch of black pepper
120ml (4fl oz) homemade nut milk
    (recipe on the right)

STEWED APPLE
apples
orange peel
cinnamon stick

TO SERVE
goat's milk yoghurt

SERVES 1
Cooking time 20 minutes, plus 1 hour to overnight chilling

Place the chia seeds into a glass and add the grated turmeric and ginger, honey and pepper. Pour in the nut milk and stir for 30 seconds, then let sit for 10 minutes and stir again. The chia seeds should be floating evenly in the milk. Place in the fridge to sit for an hour to overnight.

For the stewed apple, you need roughly one apple per serving, but make more because stewed apples last and are great at this time of year. Peel and chop the apples and place into a pot with 4 teaspoons water, 1 finger of orange peel and ½ a cinnamon stick for every 5 apples. Cook until soft over a low heat, stirring at intervals and adding extra water if needed.

To serve, allow the apples to cool and add a generous spoonful on top of the chia pudding. Here, I have also added a spoonful of goat's milk yoghurt. Play around with the toppings: winter berry compote, nuts, seeds and granola are very nice.

NUT MILK
300g (10½oz) unsalted nuts
filtered water
pinch of salt
nut bag

METHOD
Place your nuts in a bowl – I like to use a mix of almond, hazel and cashew. Cover with an inch of water, then cover with a cloth and leave overnight to soak.

Rinse the nuts and place in a blender with 8 cups of fresh filtered water and a pinch of salt, keeping the ratio one-part nuts to two-parts water. Pulse the blender a few times before blending continuously until white and opaque.

Remove and strain through a nut bag, squeezing the mixture to push through all the liquid.

Gluten-free sweet potato gnocchi with roasted pumpkin, sage & pine nuts

Gluten (for some of us) really irritates the intestinal lining and causes bloating. This is a lovely homemade gnocci option, perfect for the season as the weather starts to turn and become cooler. Pumpkin is also a great source of beta carotene, a powerful antioxidant that is converted into vitamin A.

THE GNOCCHI
750g (1lb 10oz) peeled potatoes
250g (9oz) peeled sweet potato
250g (9oz) gluten-free flour, plus extra for dusting
50g (1¾oz) tapioca flour
15g (½oz) fine sea salt
1 medium free-range egg

THE SAUCE
800g (1lb 12oz) pumpkin
2 onions
125ml (4fl oz) vegetable broth
4 sprigs of sage
25g (1oz) pine nuts
salt and pepper

TO GARNISH
8g (¼oz) pine nuts
olive oil
12 sage leaves
shavings of aged pecorino cheese

SERVES 6
Cooking time 1½ hours

METHOD
Place the potato and sweet potato in a pot of water and bring them to a boil. Remove the potatoes when they are soft – you should be able to easily push a knife through them. Place in a mouli or a sieve whilst still warm and push the potatoes through (alternatively you can grate them). If the potatoes are not smooth, sieve them again.

Place gluten-free flour, tapioca flour and salt on a flat surface and add your grated potatoes and egg. It is important to put the flour down first as this way the potatoes won't stick (too much). I always hold a bit of flour back as sometimes the dough might need a little more or a little less.

Beat the egg, before mixing into the flour and potatoes. Knead the dough until it is bouncy but not sticky. Lightly flour the surface again with gluten-free flour and roll out sections of the dough into long fingers, approximately 1cm (½ inch) wide, then cut them into 3cm (1¼ inch) lengths. Now you can use a tool or a fork to pattern them – I round the ends with my fingers. As you shape your gnocchi place them on a baking tray, making sure you flour them so they do not stick to one another.

To make the sauce, remove the pumpkin skin, cut the flesh into 2cm (¾ inch) chunks and place on parchment paper in a roasting tray. Cut the onions into eighths, leave the skin on and place in the tray with the pumpkin. Roast for approximately 25 minutes at 180°C (350°F), Gas Mark 4 and remove when soft.

Put the vegetable broth into a pot with the sage leaves, bring to a boil then simmer for a further 15 minutes before removing the sage. Remove the skin of the onions and add the broth and all the other ingredients to a blender and blend together until smooth.

Dry-fry the pine nuts for the garnish in a pan, keeping them moving so as not to burn them. Remove from the pan and add a few drops of oil. When the oil is hot, add the sage. Cook until it becomes crispy but not brown.

Bring a pot of water to the boil, salt it and add your gnocchi. When the gnocchi float to the top (approx. 2 minutes), remove them with a skimmer ladle and place them into the sauce.

Gently mix the gnocchi into the warm sauce and garnish with the toasted pine nuts, sage and shaved pecorino.

Warm radicchio salad with autumnal leaves, pear & crumbly blue cheese

This is one of my favourite autumnal salads and it's warm and hearty. Pears are full of essential antioxidants, including flavonoids, known to help fight inflammation. They are another wonderful prebiotic, helping you have a diverse and healthy gut bacteria. The bitter taste of radicchio helps activate your digestive enzymes, as well as being high in vitamin K, zinc and copper.

1 head of radicchio
olive oil
2 Conference pears
60g (2¼oz) land cress
60g (2¼oz) wild rocket
60g (2¼oz) crumbly blue cheese
60g (2¼oz) walnuts
extra-virgin olive oil
salt and pepper

SERVES 4
Cooking time 25 minutes

METHOD
Cut the radicchio into eight equal slices, first cutting it down the middle in half then halving each section again and then again. Place the radicchio onto a baking tray, drizzle with a little olive oil and bake for about 20 minutes at 180°C (350°F), Gas Mark 4 depending on the size of your radicchio head. They should be soft though and slightly brown (almost crispy) around the edges.

Whilst your radicchio is in the oven, core and dice your pears, keeping the skin on. Mix together with the salad leaves and drizzle with olive oil, adding salt and pepper to taste.

Crumble half of the blue cheese through the salad, along with half of the walnuts. Take your radicchio and place on top of the salad garnish with the remaining cheese and walnuts.

Fatima's bone broth

Fatima is the power that keeps my family home in motion: flowers appear in your room when she knows you're coming home and hearty meals upon the stove. I have long suffered with inflammatory flare-ups and the gelatine and collagen released from animal or fish bones can help soothe and repair the intestinal lining. Studies have shown collagen to be an effective remedy for people suffering with inflammatory illnesses, such as arthritis. This recipe also uses white cabbage, which is high in L-Glutamine and has been known to help repair the gut lining.

4 chicken carcasses
1 large onion
4 carrots
1 leek
1 whole celery
½ white cabbage
2 fingers of peeled fresh turmeric
2 fingers of peeled fresh root ginger
8 litres (14 pints) water

SERVES 8–10
Cooking time 4 hours +

METHOD
Place the carcass and bones in a baking tray on parchment paper and roast in the oven on a high heat for 20 minutes. This builds and deepens the flavour within the broth. If you are using bones from a Sunday roast, ignore this step. I often put my chicken carcass in the freezer, ready to be turned into broth when I need it.

Roughly chop the veg and place in a large stockpot, then add the turmeric and ginger. Pour in the water. This should cover the chicken, but the bones shouldn't be floating around; they should be snug with the veg.

Bring to a boil before reducing to a simmer. For the best gut-healing results, cook for 8 hours. The liquid will evaporate, so top it up if and when needed. You can leave in a slow cooker or on the stove over a low heat. If you're cooking it on the stove and can't stay with it for 8 hours, turn it off and pick it up again when you get back. Don't go out and leave your broth cooking! When your broth is finished cooking let it cool before sieving, holding back the bones and veg. The broth is now ready to use in other dishes, or to be simply drunk to help soothe your gut.

VARIATIONS

BONE BROTH WITH VEG
Once you have your broth, you can add anything to it. I like to add cocco bianco beans and lots of green veg. If you are using dried beans, soak them the night before, then cook them till they start to soften, throwing in your veggies at the end.

MISO BROTH
After your broth is cooked, add 2 tablespoons of miso to 1 litre (1¾ pints) of broth to make a miso broth. I like to add a little more ginger, tamari sauce and chilli too. Here I have added a soy-cured egg yolk and shiitake mushrooms.

Baked lemon & dill Norfolk trout with homemade mayo, burnt lemon & beetroot salad

This meal is light and full of flavours and textures. I don't often eat fish, so when I do I want to make sure it's giving me nourishment that I might be missing elsewhere. Trout is a freshwater fish, low in mercury and high in omega-3, which has been shown to reduce inflammation and also aids brain function. Make sure your fish is fresh and sustainably sourced. Here I have paired my trout with beets. Their wonderfully rich dark reds come from a tyrosine-derived pigment called betalain, an antioxidant that can help reduce inflammation. Beets also contain nitrate, which has been shown to reduce inflammation and lower blood pressure.

OLIVE OIL MAYO
1 egg yolk
1 teaspoon lemon juice
1 teaspoon white wine or apple cider vinegar
1 teaspoon mustard
a pinch of fine sea salt
100ml (3½fl oz) olive oil

THE FISH
3 river trout (around 350g/12oz)
4 lemons
12 sprigs of dill
olive oil
salt and pepper

BEETROOT SALAD
900g (2lb) beetroot (roughly 1 head per person)
20g (¾oz) cumin seeds
30g (1oz) mint leaves
olive oil
1 lemon

SERVES 6
Cooking time 50 minutes

METHOD
Place all the ingredients for the mayo except the olive oil into a food processor or a bowl and beat together until gently combined. Add the olive oil very slowly and continue to beat together. The mayonnaise is done when it starts to form soft peaks.

Make sure your fish have been gutted and cleaned (you can ask the fishmonger to prepare it).

Preheat the oven to 175°C (350°F), Gas Mark 4. Take one of the lemons and cut into slices, then place them in the belly of your fish along with the dill.

Place the fish on parchment paper on a baking tray, drizzle with olive oil and season with salt and pepper. Remember trout is a river fish, so it does need a little salt. Bake in the oven for 20 minutes, checking after 15. The fish should flake off the bone. Trout, like salmon, can be eaten a little undercooked, so check and decide what looks right for your taste.

Whilst your fish is cooking, cut the remaining lemons in half and grill over a medium heat on a hotplate or frying pan (do not add oil). Turn them if needed – they should be cooked through and golden.

Wash the beets and grate them. In a dry pan, toast the cumin seeds. Mix the seeds with the beets and mint leaves and massage to release the flavours, then drizzle with olive oil and the juice of the lemon. Add salt and pepper to taste.

To serve, take a sharp knife and cut along the spine/ middle of the fish from head to tail. Peel back the skin and gently remove the fillets of fish. Trout often has thin, fiddly bones, so try to avoid removing them with your fillets. Lift up the skeleton of the fish, place to one side and remove the remaining fillets.

On each plate, spread a tablespoon of mayonnaise and place two trout fillets on top. Next to the fish, place the salad and half a burnt lemon to squeeze on top of the fish.

SHROOMS

Fungi exist in a kingdom of their own, somewhere between a plant and an animal. They are the swollen womb of life and angel of death, decomposing to be reborn. Earth underfoot is brimming with sentient mycorrhizal fungi, leaping up as you pass, grabbing at debris whilst generating rich, healthy humus soil. The symbiotic relationship between flora and fungi has developed and evolved, affecting us humans more than we know, as they shape the landscapes we exist within and ecosystems we rely on. Within this underground world, there is an exchange going on, a constant trading and bartering for micronutrients, sugars, carbon and other substances. Fungi help break down micronutrients for the plants to absorb, as does our gut microbiome for us. When there are decreased fungi present in the soil, this can affect the plants' nutrient density and, in turn, we receive less nourishment from the same amount of food. Beneath the forest floor, trees tap into the mycorrhizal network, also fondly known as the Wood Wide Web, to pass information and nutrients to other trees. Suzanne Simard discovered that hub trees, or 'mother trees', tap into the fungal highway look after the younger trees by sending them their excess carbon and other nutrients they might need, depending on the time of year or their positioning within the forest. When the mother tree is dying or is injured, in a farewell bid they flush the network with a carbon message to their friends and kin.

There are 'save the Earth' types of fungi, breaking down plastics, sponging up toxic and radioactive compounds, helping habitat restoration and creating carbon-negative building materials – the possibilities are endless. Paul Stamets saw another example of the transformative effects of fungi. In an experiment, a pile of waste covered in petroleum was inoculated with fungi. Six weeks later, the fungi had absorbed the oil, broken it down and turned it into fungal sugars. When he pulled back the cover, he found large, happy, healthy oyster mushrooms, kickstarting the cycle of life. The other piles of waste were barren, yet his was an oasis teeming with life. There are also 'save your life' types of fungi, most famously penicillin, accidentally discovered by Alexander Fleming in 1928. It was considered the first natural antibiotic, although there are records of doctors working with mould since the 1800s.

Other adaptogenic mushrooms include things like cancer-fighting Turkey Tails and, my current favourite, Lion's Mane. Lion's Mane is a medicinal powerhouse, shown to improve cognitive function and related illnesses such as dementia and Alzheimer's, whilst soothing the digestive tract membrane. Then we have psilocybin, the hallucinogenic compound found in certain mushrooms, which has been seen to have had a great effect in trials to treat post-traumatic stress, depression, addiction and other psychedelic-assisted psychotherapies. I caught up with the author of *Entangled Life*, Merlin Sheldrake, to explore this wonderful web of connectivity that fungi and their fruiting bodies provide. They bear a riddle or a metaphor for us humans, as we often see ourselves as an individual, but we also are very connected to, and dependent on, the ecosystems that we live in.

A conversation with

MERLIN SHELDRAKE

Listening to Merlin's book *Entangled Life,* his words ebb and flow, interlacing scientific data into a rhapsodic enchanted tale, full of the complexities that shape the realm of fungi. Within the pages, you feel Merlin's fascination for the relationship that humans and other ecosystems have with fungal biology. Merlin recieved a PhD from Cambridge University for his work on underground fungal networks and is now a research associate of the Vrije University Amsterdam. Merlin also works with the Society for the Protection of Underground Networks and the Fungi Foundation. Connecting via the worldwide web, Merlin's voice has the same gentle tones that I became accustomed to whilst walking and listening to his fascinating compendium.

AB   Without fungi would the plants survive?

MS   No. Fungal relationships were a fundamental part of planthood many millions of years before roots, leaves, wood, flowers and fruit evolved. To this day, all plants depend on fungi in some way, although some depend on them more than others.

AB   Have fungi and plants always had a symbiotic relationship from the beginning of plants on Earth?

MS   Without fungi, the ancestors of plants would not have been able to move out of freshwater and onto the land some 500 million years ago. In the water, the algal ancestors of plants lived in a kind of nutrient soup, but on land they had to forage for minerals in a whole new way. Fungi live as mycelium and are expert foragers in this kind of environment., they were able to forage for nutrients and water on land in a way that early plants could not. Mycelium behaved like plants' root systems for tens of millions of years before plants themselves had roots. The relationship between fungi and algae gave rise to plants as we now know them.

AB   Through the psychedelic compound psilocybin, fungi allow us to connect with something that's much bigger than us, a consciousness that is greater than us. How important do you think that is in understanding — maybe not understanding consciousness because that's too big — but understanding death or becoming one with the idea of it?

MS   I think fungi can help in a few ways, not least by turning our minds towards the process of decomposition, which is how the matter and microbes that compose our bodies will continue to journey through their earthly cycles after we die. Decomposition is one of the most underappreciated miracles on the planet, and fungi are some of the most ingenious decomposers in the history of life. If nothing decomposed, the Earth would pile up kilometres deep with the bodies of animals and plants. I remember trying to wrap my head around this as a child and feeling dizzy: our lives take place in the space that decomposition leaves behind. Composers make; decomposers unmake. And unless decomposers unmake, there isn't anything that the composers can make with. In the case of psilocybin, I think it's notable that a number of recent studies into the potential therapeutic effects of psilocybin have found that it can help people deal with existential distress following a terminal diagnosis.

AB   There are other types of functional mushrooms that are becoming more mainstream, like Lion's Mane or Turkey Tail. As humans, we must have always relied on them, but it's only now that they're becoming more popular, more talked about?

MS   In East Asian cultures, mushrooms have been stable parts of medicine and gastronomy for an unknowably long time. In some sense, Europeans and North Americans are just catching up. From my point of view, the interest in medicinal mushrooms is a welcome development, although of course there is a lot of hype, as you would expect from the hyperactive wellness industry. More human clinical trials are needed, but like any research into unpatentable natural products, it's difficult to attract the necessary investment. Nonetheless, research is snowballing, and I'm

AB   Anna Boglione
MS   Merlin Sheldrake

excited to follow the space over the coming years. Fungi are metabolically ingenious and produce all sorts of potentially interesting compounds from a human perspective. They have evolved their prodigious chemical ingenuity in part because they live in intimate contact with their surroundings and are vulnerable to attack by other fungi, bacteria and viruses. They need to protect themselves, and it's often the case that when we use fungal drugs we're rehousing a fungal solution within our own bodies. Penicillin is a classic example: fungi produce penicillin to defend themselves from bacterial infection, which turns out to defend humans from bacterial infection also.

AB   There are other incredibly exciting things out there too, like fungi breaking down crude oils and glyphosate. Am I right to think that there's also a fungus that breaks down plastic?

MS   There is a fungus that has been found to break down polyurethane plastic. There may well be others.

AB   What do you think we can learn from mycelium networks in the way that they are so connected with their surroundings?

MS   I think there's a lot we can learn from fungal life to inform our systems and cultures. Fungi teach us about ecological connectivity and bridge-building: fungi are fundamentally interconnected organisms and make us more aware of the dense networks of interconnection that make life happen. Thinking about fungi gives us a vivid sense of the intricate webs of interaction and communication that we are bound up within. Fungi coax us out of our animal imaginations and human centeredness and teach us about the many ways there are to be alive: their lives are so other, and the possibilities that they face are so varied that thinking about them makes the world look different, and melts many of our cherished concepts, from individuality to intelligence. The behaviour of mycelium teaches us that a brain — and centralization — isn't required to solve problems like mazes. Decentralized solutions can be very effective responses to many of life's challenges. We are accustomed to building our societies like our bodies, but perhaps there are other ways. Fungi and their remarkable powers of decomposition might help us to revise our dysfunctional philosophies of waste. Fungi might teach us about the nature of our minds in the form of psychedelic experiences. They might also teach us about process, the open-endedness of all things in the universe. We are really patterns of stability through which matter is passing and fungi remind us of this because mycelial networks don't have a fixed shape or body, they're constantly revising and remodelling themselves. And, of course, fungi teach us to remember the power of the subvisible entities that drive so many of Earth's processes.

AB   I like the idea of everything constantly changing, I feel that leads into consciousness: if our cells are constantly dying and being regenerated, where does that leave our consciousness? It's always there, floating, while we're changing over and over, still and poised. I read something the other day that really made me think about that, how beautiful it is as a thought, that we are constantly evolving and yet we are always the same. We're always present and our mind is always here, yet it's not the same cellular structure. What's the best way to get the fungi back into the soil and what can people do at home? How can you encourage people to get more into fungi in their gardens or in their kitchens

MS   Leaving dead wood around is a good one because many fungi decompose wood. In your garden, replenishing the organic matter in the soil using compost and manure rather than chemical fertilizers, keep soil covered, stop using pesticides and fungicides… all these things will support fungal life. It's not so much a question of adding fungi to the soil. It's much better to create the conditions for healthy soil food webs to establish and then let the fungi do their thing.

Fungi don't have a fixed shape or body, they're constantly revising and remodelling themselves.

SIMON ROGAN

Arriving cloaked in darkness in Cartmel, a tiny medieval village nestled within the epic landscape of the Lake District, Atlas, my canine bestie, and I set about finding our guesthouse in this picturesque town, before turning my attention to its culinary delights. This is where Simon Rogan has immersed himself for the past twenty years, making the town his own, with three restaurants: Rogan & Co, Aulis Cartmel and L'Enclume, a triple-Michelin star eatery. Simon has expanded in Cumbria, London and Hong Kong with impressive establishments, but his heart seems to be here in this small village.

I visit him at Aulis Cartmel, his team buzzing away readying themselves for service. Simon is in the outhouse, a kitchen where dreams are made, prepping for our shoot. Simon has a great understanding of the relationship between land, produce and everything in-between. In response to Merlin's conversation, Simon has delved into the underworld, exploring the earthy autumnal vegetables and fungi that flourish in the neighbouring woods.

Dill-brined lettuce, woodland mushrooms & horseradish

We always treat the vegetables that we grow as importantly as any premium piece of meat or fish. Here we brine the lettuce as you would a nice piece of fish or meat to crisply season it and to really impart those dill flavours. It is supported by my favourite – a nice selection of wild mushrooms – all rounded off by horseradish with its spicy pungency.

10g (¼oz) fresh dill
1 litre (1¾ pints) water
50g (1¾oz) salt
4 little gem lettuces, outer leaves
  taken off
sunflower oil
240g (8½oz) wild mushrooms mix
60g (2¼oz) butter
a sprig of thyme
1 bay leaf
1 tablespoon mushroom ketchup
salt

HORSERADISH SABAYON
100g (3½oz) whole free-range eggs
  (approx. 2 eggs)
40g (1½oz) egg yolks (approx. 2 eggs)
2 teaspoons apple cider vinegar
horseradish flavour drops, to taste
160g (5½oz) butter, melted
1 teaspoon salt

SERVES 4
Cooking time 1 hour

METHOD
Mix the dill, water and salt to make a brine.

Cut the little gem lettuces in half and place in the brine for 12 hours. After this, remove from the brine and dry on paper towels, flat-side down.

In a round-bottomed bowl over a pan of simmering water, whisk the eggs, egg yolks, vinegar and horseradish drops together. Be sure the water does not touch the bottom of the bowl. Keep gently whisking and cook until it thickens (about 5 minutes). Do not overcook. The whisk should just be able to leave a slight trace in the mixture, then it is done! Remove from the heat and incorporate the melted butter and salt into the sabayon. Keep warm.

Place a pan on a high heat, add a small amount of sunflower oil and sear the little gem lettuce leaves, flat-side down, so that they begin to colour. Remove from the pan when caramelized nicely, then add the mushrooms with a little more sunflower oil. Once they begin to colour, add the butter, thyme and bay leaf and season with salt. Remove the mushrooms and discard the thyme and bay. Take a little of the cooking juices and mix with the mushroom ketchup.

Put two halves of lettuce in the centre of each plate, spoon over and around the cooking juices, then arrange the cooked mushrooms over the lettuce. Finish with the horseradish sabayon over the mushrooms.

Grilled salad smoked over embers, truffle custard & elderflower

A dish conceived from the amazing selection of kales that thrive in our farm's Cumbrian climate. Flavour is everything and this has plenty of it when you combine the kales with the truffle and the elderflower, but texture here is king. At the bottom is the soft truffle custard, made from truffles we source in Wales. Top this with the kale layers, which have varying degrees of crispness and smokiness, all cut with the floral acidity of the elderflower dressing.

TRUFFLE CUSTARD
200ml (7fl oz) whole milk
200ml (7fl oz) double (heavy) cream
1 teaspoon salt
10g (¼oz) black truffle, finely chopped
a dash of black truffle oil
90g (3¼oz) egg yolks (approx. 4 eggs)
1 free-range egg

ELDERFLOWER DRESSING
70ml (2½fl oz) elderflower vinegar
100ml (3½fl oz) rapeseed oil

KALE
binchotan charcoal
240g (8½oz) mixed kale, such as cavolo nero, red Russian, peacock, emerald ice
rapeseed oil
100g (3½oz) sunflower seeds
salt

brassica flowers, to garnish

SERVES 4
Cooking time 1 hour–1 hour 30 minutes

METHOD
To make the truffle custard, combine the milk, cream, 5g (⅛oz) of salt, chopped black truffle and truffle oil in a pan. Bring up to a simmer, then remove from the heat and leave to infuse for 20 minutes.

Whisk the egg yolks and egg together and combine with the infused liquid.

Pour the liquid into a plastic rectangle container with a lid and steam at 88°C (190°F) for approx. 20 minutes or until just cooked.

Make the elderflower dressing by mixing the elderflower vinegar, the rapeseed oil and a pinch of salt.

Light the binchotan charcoal in a BBQ and leave to go white. Dress the kale mix in a little rapeseed oil and cook over the coals until lightly charred and with varying degrees of crispness.

Lightly toast the sunflower seeds in some rapeseed oil in a non-stick pan until golden and season with salt.

Place a spoonful of the custard into the centre of a plate and arrange the cooked kale over the custard. Dress with the elderflower vinaigrette. Sprinkle over the sunflower seeds and finish with the brassica flowers.

Pan-fried Hen-of-the-Wood mushrooms, hazelnut whey, cured egg & truffle

CURED EGG
200g (7oz) fine salt
100g (3½oz) sugar
4 egg yolks

TRUFFLE PURÉE
600g (1lb 5oz) chestnut mushrooms
80g (2¾oz) butter
310g (11oz) double (heavy) cream
20g (¾oz) black truffle, shaved
2 teaspoons apple cider vinegar
salt

HAZELNUT WHEY
500g (1lb 2oz) yoghurt whey (see below)
2 tablespoons hazelnut oil
½ slice of crustless white bread
salt and pepper

HEN-OF-THE-WOOD MUSHROOMS
a splash of oil
600g (1lb 5oz) Hen-of-the-Wood mushrooms
75g (2½oz) butter
2 garlic cloves
a sprig of thyme

20g (¾oz) hazelnuts, to garnish

YOGHURT WHEY
To create yoghurt whey, place the yoghurt in a saucepan, bring it to the boil gently, then strain through a cheesecloth and discard the solids.

SERVES 4
Cooking time 1 hour, plus overnight curing

METHOD
Mix the fine salt and sugar together, cover the bottom of a tray with half of it, then carefully place the egg yolks on top. Cover with the remaining salt mixture and leave to cure for 24 hours.

Carefully wash off the salt mix and dry on an absorbent cloth, then dry a little further on a lined tray in a low oven or 60°C (140°F), Gas Mark ¼ overnight.

To make the truffle purée, slice the chestnut mushrooms and cook in a pan with the butter until soft. Add 300g (10½oz) of the double cream and reduce it by half. Blend with the truffle and cider vinegar until smooth and season with salt.

Reduce the yoghurt whey slowly in a pan to 250g (9oz). Blend the whey with the hazelnut oil, bread and remaining cream. Season and reserve.

Roast the hazelnuts in a 180°C (350°F), Gas Mark 4 oven until golden brown.

Heat a frying pan and pour in a splash of oil. Add the Hen-of-the-Wood mushrooms and season with salt. Cook until they start to go golden brown, then add the butter, garlic cloves and thyme. Baste the mushrooms steadily in the butter until it turns a nut-brown colour and the mushrooms are nicely caramelized. Remove from the pan and drain any excess fat on kitchen paper.

To plate, place a spoonful of the truffle purée in the centre of each plate. Put the mushrooms on top and then pipe a little more purée on the mushroom.

Heat the whey sauce to 80°C (176°F), then blend with a hand blender to foam. Spoon around the mushroom and finish by grating cured egg and roasted hazelnut over the top.

Miso-cooked Cylindra beetroots, smoked beet sauce & pickled mustard

I love beetroots, especially the ones that we grow for this dish – Cylindra. Here we take them and their beautiful earthiness and enhance with a glaze of miso to give them a deep umami flavour, beautifully matched with a smoked beetroot sauce spiked with elderberry vinegar and little pickled mustard seeds that explode in the mouth to provide the fresh acidic surprise.

60g white wine vinegar
40g water
20g sugar
20g mustard seeds
300g white miso paste
50g mirin
250g beetroot juice
240g Cylindra beetroots, peeled and cut
   into chunks
100ml brown chicken sauce
35ml elderberry vinegar
15g lemon juice
pinch of kuzu
red oxalis leaves

SERVES 4
Cooking time 1 hour 30 minutes, plus 12 hours pickling

METHOD
In a heavy-based saucepan, bring the white wine vinegar, water, and sugar to the boil. Put the mustard seeds into a metal bowl and pour over the boiling pickling liquid. Cover and leave in a cool place for a minimum of 12 hours.

Blend the miso paste, mirin and 150g beetroot juice until smooth. Put the beetroots into a vacuum bag with the glaze, seal and cook at 90°C (194°F) until cooked. Take the beetroots out of the bag (but keep the cooking juices) and dry in an oven for 4 hours at 60°C (140°F) to intensify their flavour a little more. Then return them to the cooking juices.

With a hand blender mix the reamining 100ml of beetroot juice with the brown chicken sauce, elderberry vinegar and lemon juice then add a pinch of kuzu. Slowly heat up stirring all the time to cook out the kuzu and for it to thicken to a nice sauce consistency. Check the seasoning.

Warm the beetroots in their glaze, then transfer to a plate and spoon on the sauce. Sprinkle over the pickled mustard seeds and some red oxalis leaves.

Jerusalem artichoke ice cream, miso caramel & malt soil

We grow an abundance of Jerusalem artichokes on our farm in Cartmel Valley. They are so versatile and can be used in many ways, including for desserts, in this case a very tasty smooth and silky ice cream. Add to that the sweet and salty miso caramel and the earthy malt and hint of cocoa soil and you get a csimple but rich decadent dessert fitting for any meal's ending.

JERUSALEM ARTICHOKE ICE CREAM
100g (3½oz) unsalted butter
200g (7oz) Jerusalem artichokes, washed and grated
300ml (10fl oz) whole milk
100ml (3½fl oz) double (heavy) cream
1 tablespoon milk powder
220g (7½oz) caster (superfine) sugar
4 egg yolks

MISO CARAMEL
70g (2½oz) sugar
110g (4oz) unsalted butter
130ml (4½fl oz) double (heavy) cream, warmed
1 teaspoon miso paste

MALT SOIL
100g (3½oz) unsalted butter
100g (3½oz) sugar
100g (3½oz) ground almonds
50g (1¾oz) plain (all-purpose) flour
15g (½oz) dark malt flour
15g (½oz) cocoa powder
2 teaspoon salt
pinch liquorice powder

YOU WILL NEED
ice-cream machine

SERVES 4
Cooking time 1 hour

METHOD
Melt the butter in a large saucepan over a medium heat. Add the washed and grated artichokes and cook for 20 minutes. Strain off any excess butter and reserve the artichokes.

Bring the milk, double cream and the milk powder to the boil, whisking constantly. Remove from the heat and blend with the cooked artichokes in a blender until smooth.

Return the artichoke mixture to a clean pan and whisk in the sugar and egg yolks. Cook over a low heat until the temperature reaches 80°C (176°F), stirring the mixture continuously. Pass through a sieve and chill in a fridge overnight. Churn the ice-cream base in an ice-cream machine until frozen and store in the freezer until ready to use.

For the miso caramel, heat the sugar in a saucepan until a dark caramel has formed. Add the butter to the caramel until melted. Combine the warmed cream and miso. Pour the caramel over the cream and miso, to taste. Whisk until smooth.

Preheat the oven to 200°C (400°F), Gas Mark 6.

For the malt soil, pulse the butter, sugar, almonds, plain flour, dark malt flour, cocoa powder, salt and liquorice powder in a blender until you achieve a crumb consistency. Place on lined baking trays and bake for 10 minutes. Whisk in a food processor, then bake for another 5 minutes. Leave to cool.

To finish, place a scoop of ice cream into a bowl and spoon over the malt soil. Spoon over the miso caramel at the table just before eating.

Berkswell pudding caramelized in birch sap & stout vinegar gel

This has become one of our best-known dishes within the restaurants, served not only in the Lake District, but also at our chef's table, Aulis, in Soho and a version is also on the menu at Roganic in Hong Kong. The combination of the buttery croissant and cheese custard base, which is then caramelised in sweet birchsap, before tangy cheese is added on top, makes this truly comforting mouthful.

BERKSWELL PUDDING
6 croissants
100g (3½oz) Berkswell cheese,
    plus extra to serve
3 eggs
300ml (½ pint) double cream
pinch salt
drizzle sunflower oil
100g (3½oz) birch sap (alternatively use
    maple syrup)

STOUT VINEGAR GEL
500g (1lb 2oz) Cumbrian stout vinegar
60g (2oz) sugar
8g (¼oz) agar-agar

SERVES 6
Cooking time 4 hours, plus 12 hours for croissants to go stale and overnight chill

METHOD
Bake the croissants in the oven at 175°C (350°F), Gas Mark 4, for 30 minutes, then leave overnight to go stale. The next day slice them in half lengthways. Place 4 of the half croissants in a 23cm (9in) baking tray then grate half the Berkswell cheese over the top. Place another 4 half croissants on top, and grate the remaining cheese onto those. Place a final layer of 4 half croissants on top.

Preheat the oven to 170°C (325°F), Gas Mark 3. Mix together the eggs, cream and salt then pour the mixture over the croissant layers. Bake for 1 hour, ideally with a second 23cm (9in) baking tray filled with water on top, to press the pudding down. Remove the tray of water and continue to cook until the core of the pudding is 72°C (161°F) (approximately 1 hour 30 minutes). Leave to cool.

To make the stout vinegar gel, place all the ingredients in a pan and bring to the boil. Once boiling whisk for 1 minute then strain into a container. Once set (approximately 40 minutes) blend the gel and pass through a fine sieve.

Once the pudding is cooled, place in the fridge with another tray on top containing something heavy (we use a bottle of oil) to press the pudding again. Let it chill overnight. The following day portion the pudding into 12 servings, then fry the individual puddings in a non-stick pan with a drizzle of sunflower oil until golden brown. Add the birch sap to the pan, allowing it to coat the pudding and reduce over the heat.

Serve two puddings per person, with a large piped dot of the stout vinegar gel and a generous amount of grated Berkswell.

WINTER

As circadian creatures, we would have risen and slept in time with the sun. As the days shorten, the sun starts to show less of herself. We might not be sleeping, but we do tend to slow down and turn inward. Winter would have provided our prehistoric selves with a hunger gap, which some argue would have supported healing within the body through a natural fasting cycle. The shorter days allow us to come together around the fire, or enjoy seasonal festivities, checking in with one another and our communities.

I spoke to Dr Françoise Wilhelmi de Toledo, a physician and fasting expert, to understand if fasting is something we should incorporate into our well-being. As the nights get longer, we tend to wind down and spend more time sleeping or, for some, trying to obtain it. Sleep is proven to be medically healing in many ways, but can it go beyond that and heal our traumas? Charlie Morley, author and lucid-dreaming expert, spoke about 'shadow work', how we can discover and heal whilst somewhere between waking and sleep state. Explorer Bruce Parry and I discuss how we have developed an individualistic mind set and if we can thrive in communities.

In this chapter I travel to Lyme Regis to see what Harriet Mansell has created in response to Françoise's words on limiting our diet and then into the heart of London to find Skye Gyngell's take on lucid dreaming. The final entry in the book had to come from Petersham Nurseries, where my thoughts and ideas first took shape. Here Ambra and Jacob conjured up recipes with community in mind.

FASTING FOR

As hunter-gatherers, we would have had periods of limited to no food intake, the body naturally using up its fat deposits, making a switch from running on glucose energy to running on ketone energy. Ketone energy is created when insulin is no longer in the body policing our fat reserves. Fat cells travel to the liver and are converted into ketones. The metabolic switch happens at around ten to fourteen hours of no food. Exercise can speed up the process, although the body shouldn't be agitated when fasting, so keep it light. Fasting has been shown to reduce inflammation, help relieve symptoms of autoimmune diseases, reduce cardiovascular issues and improve the immune system and cognitive function. There is a growing body of evidence within cancer studies showing that both intermittent and periodic fasting result in benefits ranging from the prevention to the enhanced treatment of diseases. An experiment led by Valter Longo showed that mice who had undergone chemotherapy whilst fasting saw their white blood cell count bounce back to normal after the treatment. When the body goes into 'fasting mode', it starts to create stem cells. When we go back to our 'normal state', these stem cells are activated and rejuvenate damaged cells within the

LONGEVITY

body. As we age, we accumulate damaged cells, often affecting our memory. Challenging our brain's normal state, by putting it in a 'stressful situation' such as endurance exercise or fasting, improves its cognitive function by activating its adaptive stress responses. Nerve cells become more active and neurochemicals are released, creating the production of new neurones from stem cells. What counts as fasting? There are different ways to fast: some people only eat within a certain timeframe, which is called intermittent fasting. Others prefer to have a restricted diet for certain days of the week, often only consuming 500 calories a day, like the 5–2 diet, where you have a limit on the calories you consume for two days a week. Another great study on mice showed that when the mice practised intermittent fasting and ate within an eight-hour window, they lost up to twenty-eight percent of their body weight. Longer fasts might also use a restricted diet or be focused on liquid intake. Either way, you should seek advice from a medical professional as it is important to take into account your health, age and weight before starting a longer fast, as well as easing in and out of the fast.

A conversation with

DR FRANÇOISE WILHELMI DE TOLEDO

The first time I spoke to Françoise, I was in Devon struggling with a signal, in-between picking up Jeremy Lee to shoot for another chapter. Françoise has an energy that invites you to sink into the conversation at hand. We immediately started chatting, spider webbing from the original topic. Françoise's Wikipedia page will tell you that she is a 'fasting expert' and managing director of Buchinger Wilhelmi, a clinic founded by her husband's grandfather. Buchinger boasts a hundred years of therapeutic fasting. They have been at the forefront of clinical research and Françoise and her team are constantly integrating new knowledge into their clients' experiences, creating a deeper level of healing with every visit.

AB   Naturally, we should be fasting because we are hunter-gatherers. Hunter-gatherers don't have food on tap the whole time; we have to go out and look for it. We release dopamine to encourage us, which now we release too easily, so we overeat and without pauses.

FWdT Your point on gatherers and hunters is totally true. There were long seasonal periods of time where humans had no food, and they had to rely on their own caloric reserves: in a time when people had no technology to conserve food, the only place where they could store food was in their own body, as fat. So, the cosmic rhythm of the Earth moving around the sun in one year triggers the seasons. In seasons with low sun exposure most animals either migrate to other places where they find sun and food or have longer periods of fasting and hibernate. Some scientists think that humans hibernated too. The reason why we accumulate fat is because of these long gaps in the availability of food, for instance in the wintertime. Another short cosmic rhythm comes from the revolution of the earth on itself lasting twenty-four hours and causing the alternation of darkness and daylight. At night, we have to sleep. Sleep, fast and darkness are programmed in our circadian brain and our circadian clocks within our organs. Again, we should fast during the nighttime and over seasons of low sun exposure and eat and be active during daytime and the seasons of high sun exposure. We have thus two nutritional programmes. 'Nutritional programme one' is eating and 'nutritional programme two' is fasting.

AB   Am I right to think that when you go into starvation mode, you create stem cells and then when you come out of it, the stem cells are activated and they go and heal damaged cells?

FWdT Well, I don't want you to use the word 'starvation' here because starvation only happens when you have no fat reserves anymore. For a person of a normal weight starvation generally only occurs after around forty days without food!

AB   My mistake.

FWdT Furthermore, when you are obese, you can fast for over 300 days before starvation is reached. One publication reported the longest human fast of 382 days in a massively obese person. So, let us use the word fasting for the usual periods of five, ten, fifteen, twenty days and sometimes more. Accounting for your age, your physical activity and the size of the meal you had before, you need ten to twenty-four hours to reach the metabolic switch, the change of programme from eating mode to the fasting mode. Switching metabolism means switching fuels from carbohydrates or glucose to fats and ketones. Ketones are derived from fat, the fat of your own stores, in the adipose tissues of the body.

This switch triggers a lot of changes. In the brain, BDNF (brain-derived neurotrophic factor) is produced, which sharpens the senses, enhances alertness and activates neuronal regeneration out of stem cells. These stem cells that are stimulated during fasting are released when food is reintroduced. It 'rejuvenates' to a certain extent the tissues, enhancing the functions

AB   Anna Boglione
FWdT Dr Françoise Wilhelmi de Toledo

of apoptosis and autophagy. Some tissues like the digestive organs regenerate their cells in few days, others in few weeks or months.

AB It's quite amazing, the power of the body, that just through something as simple as taking away your food, the body can heal itself. It's something that everybody can do. Obviously, if you're doing a long fast, you need to be in a controlled environment. If people are thinking about fasting, what would you say to them? Who should be doing a fast?

FWdT It's much more normal to fast some hours every night and for a longer period every year than not to fast at all. So, in my opinion everybody should fast. The question is how long and with which frequency, and with or without supplements? If you are healthy, you do it in a preventive manner to stay healthy as long as possible, thus retarding or avoiding the emergence of diseases linked to aging. Daily intermittent fasting is just something we are programmed for, between ten and sixteen hours without eating at all. This is part of our physiology, not some special diet. If you are overweight, try to fast sixteen hours overnight. If you're a normal weight, then you have to find the length of daily fasting that fits your lifestyle.

A longer period of fasting, once a year, I think is necessary for everyone. If you're ill, then there are special rules and groups for whom fasting would be particularly beneficial.

The first group is those with dysmetabolism, which means being overweight and having abdominal obesity, linked most of the time with insulin resistance, high LDL, high triglyceride and high blood pressure. This makes you prone to cardiovascular accidents like myocardial infarction and stroke as well as diabetes type II.

The second big group are all the chronic inflammatory diseases like arthritis or allergies, asthma or migraines. All conditions that are painful, red, warm and swollen suggest inflammation. The length and type of fasting should be dependent on the biological age of the person and their nutritional status. In our clinics the doctors are the ones who determine this. If someone is already very thin, for

of the organs. So, while fasting is the 'cleaning part' of the process, where you get rid of all damaged and senescent cells, the food reintroduction after the fasting replaces the eliminated cells by young ones.

AB There have been studies on mice with cancer showing that fasting whilst doing chemotherapy allows the white blood cells to bounce back. Is this true also in humans?

FWdT It is true in humans and all animals. In the case of cancer and chemotherapy, all the immune cell lines will tend to decrease during a fasting period. At the same time, stem cells are activated in the bone marrow and on food reintroduction mature immune cells will be released, thus enhancing immunity. After some days, the pool of immune cells is reconstituted, but in a rejuvenated way. In other words, the cell count diminishes by getting rid of the old and damaged cells, then re-expand with fresh young ones. This phenomenon of cell regeneration happens in all body tissues and is called 'cellular turnover'. Fasting accelerates it considerably using the mechanisms example a woman with polyarthritis, we might have to find other nutritional strategies in order not to deplete that person too much.

The third group includes those with a depressive mood, pre-burn-out or physical and emotional exhaustion. Fasting gives an emotional and physical energy boost.

The fourth group includes the severe diseases such as Parkinson's and epilepsy. It is well-known that fasting and a very restrictive ketogenic diet that excludes carbohydrates to different extents can eliminate seizures in epileptic children who cannot use drugs. There are many studies on multiple sclerosis and the prevention of Alzheimer's. Fasting has certainly anti-tumoral effects. A great deal of research on fasting periods or dietary interventions to treat cancer and reduce side effects of chemotherapy is going on.

In case of illness, fasting has to be done in a setting with professional care. Of course, the older you are, the more supplemented this type of fasting might be. In some cases, it shouldn't be too long.

AB   What about fasting to improve mood?

FWdT For mood disorders, fasting can bring benefits. Cognitive improvement has been linked to the production of BDNF (brain-derived neurotrophic factor); it seems to enhance alertness and focus as well as neuronal regeneration. The explanation is that in the course of human evolution the problem was generally not an excess of food, the problem was a lack of food. Animals and humans were always busy looking for it, and when they didn't find food for a certain time their survival was endangered. Thus, it was important to stay alert in order to soon refill our calorie reserves. There was no obesity in prehistoric times! This is probably the reason why the fasting metabolism is linked with an enhancement of energy, acuity and reactivity that lifts the mood.

In our clinics, doctors have observed that after some days of fasting people start to calm down, are more focused, more present, their senses are sharper, they enjoy nature, meeting others and life in general. Some people come and write a book or a film during the time that they are at the clinic. We have lots of artists recovering from their stressful lives. In general, this means creativity, intuitive intelligence, is enhanced.

AB   In a way, you remove all the clutter from your brain.

FWdT I think you can put it that way. Your mental activity, the blah blah in your head, and also the sorrows and dramas, they are put aside or are less active. You can connect better to other people, to the beauty of nature when you walk in a forest, and finally to your own self. You know, in religions fasting times are times to reflect on what happened during the year, and in both the Jewish and the Christian traditions to ask for forgiveness for the errors you committed. You would try to put yourself back on the right track, which would be God's programme for you. Sometimes we are trapped in toxic situations we are afraid to leave. We are afraid of the unknown. It needs courage to step out of it and a fasting period can give you that courage.

HARRIET MANSELL

Harriet came to me through her agent Lucy. However, we soon realized that connections ran between us. My brother Harry, being a local Devonshire lad, supplies Harriet's restaurants and they have also built a friendship. I can see why; Harriet has this salt of the Earth feel to her, coupled with a strength that has seen her open two restaurants within a very short space of time. She explains to me that she never meant to open Lilac, a small plates restaurant and wine bar. Originally it was meant to be storage space for her restaurant Robin Wylde. As soon as she got the keys, ideas started to flow.

Her food is seasonal, rustic and refined. I know it seems odd to put rustic and refined in the same sentence, but her dishes include contrasting elements, ones that create nostalgic memories of home cooking and others that incorporate the graceful techniques of a well-trained chef. If you make the trip to Britain's Jurassic Coast, it's well worth a stop in Lyme Regis to try both Robin Wylde and Lilac.

Warming broth with seaweed

Fasting often refers to consuming a limited number of calories, but on fasting days, it is still vitally important to warm and nourish the body. Warming, andthereby resting, the body is something that also helps us to engage with our parasympathetic nervous system – essential for healthy homeostasis. Having something warming that is full of probiotics and prebiotics, whilst harnessing the nutrients of local and seasonal produce, and being a liquid that is easy to process – as well as combining the ritual of mindfully making a broth – make this essential for those times we wish to place our body in fasting mode.

a handful of dried seaweed (we use kelp)
10 chestnut mushrooms
a handful of dried mushrooms
3 onions
3 garlic cloves
1 fennel (if in season, it's a nice addition)
1 carrot
1 leek
7.5cm (3 inch) piece of fresh root ginger
4 celery sticks
6 star anise
a couple of sprigs of thyme or oregano
2 litres (3½ pints) water
1–2 tablespoons fermented fava bean paste
   or brown miso (barley or rice), to taste
fresh seaweed, to serve (optional)
salt and pepper

SERVES 6
Cooking time 1 hour

METHOD
Peel, trim and roughly chop the vegetables as required, then put the first 12 ingredients into a pot with the water and bean paste or miso and bring to the boil. Reduce the heat to simmer for 1 minute before turning the heat off and leaving to infuse and cool.

Reheat and season to serve. Fresh seaweed to serve can be a tasty mineral addition, but no need.

NOTE
In the restaurant we serve this with a yeast oil (made from fresh baker's yeast, which we caramelize in the oven at 180°C (350°F), Gas Mark 4 before blitzing into an oil with some salt), then drizzle into the bowls. This really makes for a full-blown warming umami hug in a bowl, tantalizing the taste buds.

Cuttlefish with cabbage

Cuttlefish is an often overlooked ingredient, which I like to use when it comes in as a by-catch. The texture is firm, and it is a very good source of lean protein. Fasting days can be focused on consuming low calories, often under 500. This is a small, low-calorie meal, packed full of flavour and texture, designed to satisfy and really fill you up without the inclusion of carbohydrates. It makes a good choice of meal for those looking to take their bodies into ketosis, whilst also feeling that all their taste buds have been satisfied.

1.5kg (3lb 5oz) cuttlefish
2 onions
a splash of light olive oil
4 garlic cloves
5cm (2 inch) piece of fresh root ginger
200g (7oz) butter
sweetheart cabbage
oil of your choice, such as olive oil or smoked
    rapeseed oil (we use yeast oil), for the cabbage
a squeeze of lemon juice
sea salt flakes

LARCH-INFUSED VINEGAR
handful foraged young larch or pine tips,
    alternatively use a few sprigs of thyme
    or rosemary
apple cider vinegar

SERVES 4–6
Cooking time 1–1.5 hours, plus 1 week for infusing vinegar

METHOD
For the larch-infused vinegar, we take larch or pine tips and sit them in cold apple cider vinegar, allowing to steep for a week or so in order that the vinegar takes on the flavour of the larch. This adds an aromatic, citrusy and resinous flavour to the vinegar, which is then used to season the dish.

Prepare your cuttlefish by removing the skin, cleaning out the ink and pulling away the membrane. Using a sharp knife, slice it very thinly – we are talking the thinnest you can get it, a little like fettuccine. Set aside in the fridge.

Peel and slice the onions lengthways and add to a frying pan with a splash of olive oil and a pinch of salt. Sauté gently – you want to soften them, not brown them. Peel and slice the garlic and add to the pan to soften with the onions. Peel the ginger, finely dice or mince, and add to the onions. Once softened, take off the heat and leave to one side.

In a separate pan, put the butter and put over a medium heat to darken to make a brown butter (beurre noisette). When brown and

smelling nutty, after it has frothed up and turned a nice medium-brown colour, remove from the heat, strain and cool.

Preheat the oven to 180°C (350°F), Gas Mark 4.

Remove the leaves from the pointed cabbage, wash and steam briefly. Lightly coat in the oil of your choice, place on a baking tray and bake for 6–8 minutes to crisp up a little bit.

When ready to serve, add the brown butter to the onion mixture with a tablespoon of larch vinegar, a squeeze of lemon juice and a big pinch of sea salt.

Add the cuttlefish and cook over a medium heat and stir continuously until just cooked through, about 20 seconds to 1 minute – this shouldn't take much longer. When just cooked through (be careful not to overcook), taste and possibly add more larch vinegar and salt. It benefits from really popping with flavour to contrast with the earthy cabbage leaves.

To serve, cover a plate with the cabbage leaves and sprinkle over extra salt flakes. Pile the onions and cuttlefish on top.

Oat & seed 'porridge' loaf

This loaf is packed full of protein and is designed to really fill you up for a long period of time, whilst delivering essential nutrients. You can eat a slice of this by itself as a snack, for breakfast or to accompany a soup and it will really satisfy hunger and deliver on nourishment, while offering a slow release of energy.

NOTE
Use a 500g (1lb 2oz) yoghurt pot
   as your measure

1 pot of plain yoghurt
1½ pots of mixed seeds
   of your choice
2 pots of oats
3 pots of eggs
1 tablespoon bicarbonate of soda
   (baking soda)
1 teaspoon table salt
a good splash of milk, plus more
   to adjust the consistency

MAKES 2 LOAVES, SERVES UP TO 10
Cooking time 1 hour

METHOD
Preheat the oven to 180°C (350°F), Gas Mark 4.

Find a large bowl, add all the ingredients and stir together. You need to add enough milk to create a porridge-like consistency. This recipe is fairly forgiving, so no need to worry if it's too thick or not thick enough – just go for a porridge-like consistency.

Allow it to sit for a couple of minutes for the oats to absorb the liquid and possibly then adjust the consistency further with another dash of milk.

Line two small loaf tins with parchment paper and divide the mix between the two tins.

Bake in the oven for 1 hour or until the loaves are firm to the touch.

Yoghurt sorbet, fig-leaf oil & mulled grape jelly

This is a recipe for those who want a little something sweet or want to soothe their digestion. The dish in its entirety is designed to calm digestion and be cooling. Lightly sweetened yoghurt sorbet, combined with the relaxing fig-leaf oil and gentle spices in the jelly, make it an ideal low-calorie dessert or something to ease you into or out of a fast.

SORBET
1kg (2lb 4oz) plain yoghurt
150g (5½oz) sugar
150ml (5fl oz) water

FIG-LEAF OIL
a handful of fig leaves
500ml (17fl oz) light olive oil
a pinch of sea salt flakes

MULLED GRAPE JELLY
800ml (1½ pints) grape juice
a handful of mulling spices (can include
    cinnamon stick, allspice berries, whole cloves,
    cardamom pods, star anise, bay leaves)
200g (7oz) sugar
8 gelatine leaves

SERVES 10–12
Cooking time 10 minutes, plus 30–60 minutes churning time

METHOD
In the restaurant, we put all the ingredients for the sorbet into a Paco beaker and set, then churn. However, at home, you could place the ingredients in a countertop ice-cream machine. Churn until set, approximately 30–60 minutes in a home machine, then place in the freezer. For the best consistency when eating, churn on the day you are serving and set for no more than a couple of hours. Otherwise remove it from the freezer and allow it to soften a little prior to serving.

To make the vibrant green fig-leaf oil, we use a Thermomix. Roughly chop the leaves a little before putting in the blender. Add the oil and salt, and blitz on the highest speed for 30 seconds. We then heat while blending on speed 6 to 82°C (180°F), before finishing on speed 10 for a further 1 minute. Strain, rapidly cool and place in the refrigerator to preserve the colour and flavour.

If you do not have the privilege of having a Thermomix at home, you could use a jug blender and use your judgement or a thermometer to gauge the temperature, then remove from the blender and cool and strain.

To make the mulled grape jelly, heat the grape juice with the mulling spices to infuse, without bringing to the boil, and add the sugar. Leave to infuse until cool.

Bloom the gelatine by placing the sheets in a jug or bowl of water for around 1 minutes, or until the consistency begins to soften and become jelly-like. Remove the squishy ball of gelatine from the water and squeeze out the excess water. Add this to the juice, stir in, then pass the juice through a sieve into a baking tray to set.

To serve, spoon some of the jelly into a bowl, next to some sorbet, and squeeze some of the fig-leaf oil around the bowl.

Miso-glazed squash with smoked quinoa & wild pickles

A hearty, healthy and nutritionally complete meal packed full of flavour. We use English smoked quinoa – so tasty and a complete source of protein – combined with glazed squash and an acidic hit from the pickles. This acts as a low-calorie but dense, filling and exciting meal to get you through a winter fast, or a low-calorie day when you need a full meal. This would also be a good meal the day after a fast.

2 tablespoons brown miso
2 tablespoons sesame oil
4 garlic cloves, minced
5cm (2 inch) piece of fresh root ginger, minced
1.5kg (3lb 5oz) squash (any variety of your choice)
180g (6½oz) smoked quinoa (we use English Hodmedod's)
juice of 1 lemon
100ml (3½fl oz) light olive oil
a handful of mixed seeds
a small jar or so of wild garlic pickles (see Note)
salt and pepper

SERVES 6
Cooking time 1 hour

METHOD
Preheat the oven to 180°C (350°F), Gas Mark 4. Mix the miso, sesame oil, minced garlic and minced ginger in a bowl.

Slice your squash into wedges and coat in the miso mix. Place on a baking sheet and bake in the oven for 20–30 minutes until the squash is tender and the outside is caramelized.

Boil the quinoa for 10 minutes. Drain and season with lemon juice to taste, the olive oil and salt and pepper.

Toast the seeds for 7–10 minutes, then season. Add to the quinoa mix.

Divide the quinoa between six plates, topping with the squash and wild pickles.

NOTE
We made the wild garlic pickles earlier in the year. If you don't have these, look for some other tasty pickles of your choice that could liven up the plate and add a zing.

Mushrooms, Staffordshire oatcake, goat's curd, chicory & lemon dressing

A small dish that ticks off each taste sensation – sweet, savoury, bitter, umami, salty. We use a thin oatcake, much like a crepe or a chapati, as a base for earthy, umami mushrooms, topped with bitter leaves dressed in an acidic and salty dressing. It is a simple, delicious, small and beautiful meal, which will satisfy and fill you up whilst offering deliciousness and a multitude of nutritional benefits. The lemon verbena dressing can of course be swapped out for a citrus dressing, but if you can find some fresh lemon verbena, this will really be the aromatic finale this little dish needs, as well as being something so good for the digestion. If you can forage for your mushrooms, this will also make this little dish next-level.

STAFFORDSHIRE OATCAKES
(We use Felicity Cloake's recipe for the oatcakes, which makes 10 large, so there are a few spare)

450ml (16fl oz) milk
450ml (16fl oz) water
250g (9oz) finely ground oats (you can grind ordinary oats in a food processor)
100g (3½oz) strong wholemeal flour
100g (3½oz) strong white flour
1 teaspoon fine salt
4g (⅛oz) dry yeast or 10g (¼oz) fresh yeast
fat of your choice (lard, bacon drippings, clarified butter, vegetable oil)

MUSHROOM MIX
750g (1lb 10oz) mushrooms of your choice (see Note)
2 tablespoons cream cheese
a squeeze of lemon juice
1–2 garlic cloves, minced
50g (1¾oz) dried mushroom powder
sea salt and pepper

CHICORY SALAD
100ml (3½fl oz) lemon verbena oil (see Note) or light olive oil
100ml (3½fl oz) lemon juice
a good teaspoon salt
500g (1lb 2oz) chicory (see Note)
250g (9oz) goat's curd

SERVES 6
Cooking time 45 minutes–1 hour

METHOD
To make the oatcakes, heat the milk with the same amount of water in a pan until about blood temperature.

Meanwhile, mix the oats, flours and salt in a large bowl. Mix the yeast with a little of the warm liquid and then cover and leave until frothy. Stir into the dry ingredients then whisk in the remaining liquid until smooth. Cover and leave in a warm place for about an hour until bubbly, or overnight in the fridge if you prefer.

Grease a large frying pan with a little fat and put on a medium–high heat. When hot enough for the batter to sizzle as it hits it, give the bowl a quick whisk, then add a ladleful to the pan and tilt to spread it out. Cook until dry on top, then loosen the edges and carefully turn it over. Remove from the pan when cooked on both sides, then repeat with the remaining batter to make 9 further oatcakes.

To make the mushroom mix, in a dry frying pan cook the mushrooms on their own for a minute or two to remove any moisture, then add a couple of tablespoons of cream cheese, a splash of lemon juice, the garlic, dried mushroom powder and black pepper and sea salt flakes to season. Cook until it all comes together and adjust the seasoning to suit your tastes.

Make the dressing by combining equal parts lemon verbena or olive oil to lemon juice with the salt and shake or whisk it together. Pull the delicate outer leaves of chicory in petal-type shapes from the white central 'spine', which is a little bit more bitter. Dress these, possibly adding a few extra sea salt flakes, if needed. Adjust the lemon juice level. too if needed, depending on how bitter the leaf is.

To serve, place an oatcake on each plate, add the hot mushrooms, crumble over some goat's curd or light goat's cheese and cover with the dressed chicory leaves.

NOTE
We use Rosalba and Fenice chicory as these are available from our local farm; however any bitter leaves would be a nice contrast here. The idea is that you can use very bitter leaves and season them to contrast the earthy mushrooms and goat's cheese.

We used chanterelles and hedgehog mushrooms because they were available at the time of shooting the recipe. You could substitute for any other mushrooms at any other time of year, including shiitakes, chestnut mushrooms, oyster mushrooms or whatever you can find.

If you want to make lemon verbena oil, simply blitz lemon verbena with light olive oil and heat to around 80°C (176°F) before cooling.

DREAM

bodies has a circadian rhythm, including our heart rate, blood pressure, melatonin, cortisol and triglycerides. Two external factors that are known to be very disruptive are food and light. When our circadian rhythm is off-kilter, this can contribute to a variety of illnesses and ailments. The World Health Organization goes as far as classifying night shift work as a carcinogen, due to its disturbance of our sleep cycles. According to sleep expert and author of *Why We Sleep*, Matt Walker, anything below eight hours can affect our health and erode our DNA code. Surviving on only six hours of sleep sees our immune function decrease and the DNA that controls tumour growth, inflammation and stress all increase. Four hours of sleep a night sees a seventy percent reduction in natural killer cells, which are vital to fight off cancer and viruses.

For some of us, sleep is more obtainable than for others; however, it is near impossible to move through life without having one of those restless nights. Thoughts might creep in: you've forgotten to do something, your pillow is no longer comfortable, your bed is suddenly hard and no amount of tossing and turning will get you to sleep. As children we have a rhythm, kindly enforced by our guardians. We eat at certain times of the day, we go to sleep when the sun goes down and (often) wake when it comes up. This all helps calibrate our internal twenty-four-hour clock, known as our circadian rhythm. The word derives from the Latin 'circa' meaning 'around' and 'diēm' meaning 'day' – one day round the sun. Every cell and function in our

LAND

Sleep also allows us to commit what we learn and new experiences to memory. The saying 'sleep on it' refers to one's creative problem-solving abilities in the state of REM. What else happens when we slip into dream land? In my next interview, I wanted to explore another element of sleep, when we are neither in a waking state nor dream state. I knew the person to talk to, to better understand the topic of Dream Yoga and how we can improve ourselves and heal our trauma in the dream world. Charlie Morley is a lucid-dreaming teacher and author of five books (and counting) on the subject. We spend, on average, twenty-five years of our lives sleeping. A lot of this time is spent in 'ignorance', without a connection to what is happening. What if we can change that to better understand ourselves, clear blockages, discover and enhance our inner qualities?

A conversation with

CHARLIE MORLEY

Charlie has been lucid dreaming for over twenty years. He is an award-winning author and has given many talks on the subject, including for TED and Oxford and Cambridge Universities, and was 'authorised to teach' within the Kagyu School of Tibetan Buddhism by Lama Yeshe Rinpoche. As well as teaching workshops on lucid dreaming, Charlie also works with veterans to release post-traumatic stress.

I know Charlie from him hanging around our house when I was still a kid. Always dressing in blue, to the point he was known as 'Charlie Blue', he would cover topics from hip-hop to Buddhism. He was always my older sister's partner-in-crime, and he immediately breaks into those stories when our faces pop up on one another's screens. An experience he had with my sister, which would later linger, pushed him into lucid dreaming to heal trauma instead of, as he puts it, 'having fun flying around, skateboarding and having sex with celebrities'.

AB   Anna Boglione
CM   Charlie Morley

AB   Within the body, we hold so much trauma: emotional trauma, trauma from childhood, trauma from work and life experience. You see this come out in different elements, especially within the digestive system or within the body, through aches and pains. People go to doctors and often they don't get healed. What really interested me when I was flicking through your book, *Dreaming through Darkness*, is confronting what you call the shadows. You talk about trying to fulfil this idea of self-love and working on traumas and healing them with compassion.

CM   Absolutely. With some of my most powerful lucid dreams, I've woken up and my body's shaking, but in a good way. It's like there's a release. All those years ago, I didn't know what it was, but now that I've done a lot of somatic body work and I've done the breathwork teacher training I realise, wow, it's all leading to the same point. The lucid dream is integrating a trauma to such an extent that the body is literally shaking or vibrating when you wake up. As for the shadow work, this was an idea pioneered by Carl Jung, the shadow being anything we have rejected, denied and disowned, the dark side of our psyche. The shadow is the part of me that's made up of my trauma, that's made up of my fears, made of my phobias, my prejudices, but also all the bright stuff that I hide from others. If you hide it, it's in your shadow.

We hide our trauma and we hide our fear, but we also can hide our talents and our sexuality or whatever it is. So, in the waking state you can engage certain psychological modalities to integrate the shadow and they work brilliantly. However, nothing is more visceral than meeting the shadow in a lucid dream. It all boils down to this: in a lucid dream, psychological concepts become personified. So, the lucid dream is a symbolic representation of the internal environment of your own psychology. What else would it be? You're literally in your mind. So, everything is symbolized. If you meet your mum in a dream, lucid or not, it's not your mom, it's your symbolic representation of your mom. So, if your mum appears in the dream and she's really, really angry, you can be like, 'OK, so my relationship to my mum at the moment, my inner representation of my mum, is one of anger.' Or, if I meet my dad in the dream and he's like super-wounded with cuts all over his face, I'd wake up and go, 'Oh, what part of my inner father is wounded right now?' This is a non-lucid dream. In a lucid dream, once you can actually direct the dream, you can literally become lucid and call out 'inner child, come to me' or 'sexual trauma, come to me' or 'I'm ready to heal my wounds'. The symbolic representation of child abuse will appear, or the symbolic representation of combat zone trauma will appear, that you can actually dialogue with. Not like in a shamanic journey, where you might kind of see it; not like in hypnotherapy, where you might feel it, but you're being guided by another. I mean, it will be as real as this – I actually shook the hand of 'repressed capacity for violence'. I called out for the shadow element and he appeared in this really scary form. I was like 'shadow, come to me', At the time I did a lot of martial arts training, so I think the part of me that I was hiding from others and myself was probably the repressed capacity for violence. This kind of monster appears with a face of a pig, and I go 'are you my shadow?' And he goes, 'I am your repressed capacity for violence. You will never defeat me.' I was like, 'This is insane. You're in my head. I know my body's asleep in bed. I know I'm dreaming. I know this isn't real, that it's an internal projection of my own mind.' But how specific? I could meet my shadow self, my repressed capacity for violence, and then I can literally embrace them, or dialogue with them or put your hand on their heart and kind of integrate the energy. It's like the stuff you see in movies: the symbolic representation of facing your fears. It's like that, but it's real.

AB   And when you meet your shadow, in various different forms, you show it love and kindness?

CM   Yeah.

AB   So, you're essentially showing yourself love and kindness, which is something that we don't do so much. Everybody has the negative voice in their head. Like you were saying earlier, you suppress your qualities out of fear of failure or out of fear of being seen. So, it's very interesting to then give yourself, in many different ways and shapes, that hug and that love that you might have missed out on as a child, or that you don't know how to give yourself but can give to others. In the waking state, when you do acts of kindness you release oxytocin and I'm wondering in lucid dreaming whether doing acts of kindness towards yourself also releases oxytocin?

CM   Yes. So, once you become lucid, a big brain network switches on: the prefrontal cortex. Once the prefrontal cortex switches on, your brain thinks you're awake. It doesn't think you're having a lucid dream because wakefulness for the brain is not predicated on having your eyes open. Wakefulness for the brain is predicated on activation of the prefrontal cortex. This is why if in the lucid dream you face a fear, you integrate a trauma or you heal your inner child, then it's not like a hypnotherapy session or imagination therapy or creative writing therapy. From the brain's point of view, you've actually integrated the trauma because the brain thinks you're awake. Neurotransmitters and neurochemicals are released in exactly the same way as when you're awake. Now, in June we're running one of the first scientific studies to prove that, because at the moment this is hypothetical, but we're doing a saliva study with veterans. Our hypothesis is that healing trauma in a lucid dream is so visceral, you'll have anti-inflammatory biomarkers appear in the bloodstream or the saliva after a lucid dream healing. If we can prove that, and we are almost certain we will, not only will it be one of the most exciting studies for lucid dreaming, but it will also prove the mind-body connection because a lot of the argument against the mind-body connection is outside influence. In a lucid dream, you're in a closed environment.

In the lucid dream you face a fear, you integrate a trauma.

It's a closed circuit, with no outside influence. So, if you can prove that the mind can affect the body in a lucid dream, it's a really pure test, a pure example of how the mind can affect the body.

AB   Yeah, that would be incredibly interesting. Going back to having ailments within the body that you can't heal, especially inflammatory issues. My little sister who you know, Ruby, her ileocecal valve is super-inflamed and super-painful and nobody can get to the bottom of it. She internalizes everything, so something like lucid dreaming could be amazing for her because she could then, maybe even without knowing what the root cause of it is, start to do some healing on herself.

CM   Yes, that's really important you said that. Healing can occur without cognitive awareness of what the trauma originally was; that's the basis of body-based healing. The new view on trauma is so different to what it was even ten years ago. You don't need cognitive awareness of what the trauma was to integrate it through the body. Which is huge. This completely overturns the Western psychological model of talk therapy, which is basically that if you talk about it enough, you'll integrate the trauma. But trauma is held in a part of the body that doesn't respond to words. So, you can talk all you like, and therapy is brilliant – I'm in therapy at the moment – but talking therapy for trauma is pretty ineffective.

AB   I feel that in my therapy. It took me a long time to find the right therapist and I have to say that I didn't enjoy a lot of therapy in the past, but now I really look forward to it because I find the human psyche so interesting, to unlock and understand all these different threads. When you have the spider web of things that are connected and influencing each other, it is fascinating. No matter how much I verbalize the things that are in my mind and have a great understanding of them, I cannot let go. They're in my body: the pain, the suffering, probably a lot of the stomach issues that I have are all still trapped inside me. I was thinking, as I was talking to my therapist, asking her how I let go of certain things, that I should delve into one of your books or come to your retreats to learn to heal from the depth of my psyche.

SKYE GYNGELL

I remember the excitement, the ideas, the flavours that came with the set-up of the Café at Petersham Nurseries, my mother, Skye and Lucy Grey plotting. All dishes were to roll with the seasons and Lucy would be in charge of growing and evolving the kitchen garden. The transformation began with a simple garden centre with concrete floors, ripped up and replaced with my father's prized hoggin, plants jutting up through the floor and overhead. Skye would soon enough be winning a Michelin star. There was no need for white tablecloths and fancy lighting: her food, elegant to see and to taste, spoke for itself.

An Australian trained within the French culinary tradition, adopted by the British and acclaimed by all who eat her food, nowadays you can find Skye at the helm of her restaurant Spring, located in Somerset House on the Embankment. It is beautiful, simple and refined in décor, with sweeping light and dashes of pink. When thinking of this book, naturally Skye came to mind. She has in some way provided the narrative to my understanding of sweet, sour, salty, bitter and umami.

Cima di rapa with agresto

Nuts, such as almonds, walnuts, pistachios and cashews, contain a combination of melatonin, zinc and magnesium. This combination is known to help aid sleep in adults. This dish also incorporates anchovy fillets, high in omega oils, and cima di rapa, a source of vitamins A, C and K, as well as potassium, calcium, iron and fibre.

AGRESTO
20 young wet walnuts in their shell
2 anchovy fillets
1 garlic clove, peeled
1 dried red chilli, finely chopped
zest and juice of 1 unwaxed lemon
200ml (7fl oz) extra-virgin olive oil

CIMA DI RAPA
a large bunch of cima di rapa
a drizzle of extra-virgin olive oil
a few drops of lemon juice
sea salt and pepper

SERVES 4–6
Cooking time 45 mins

METHOD
Crack open the walnuts and extract the nuts. Place in a pestle and mortar and pound gently. Add the anchovies and garlic and continue to pound. The sauce should be smooth and creamy in parts and rough in others.

Stir in the chilli, lemon zest and juice, then gradually pour in the olive oil. Taste and adjust as necessary – you may need a pinch of salt.

Set aside while you cook the cima di rapa. Prepare the cima as you would broccoli. You want to use all of the vegetable – the leaves, flowers, stems and stalks.

Place a large pot of well-salted water on to boil. Once the water is boiling, plunge in the cima. Allow the water to come back to the boil and cook for 2 minutes – the cima should be tender to the bite.

Drain and drizzle with olive oil. Add a couple of drops of lemon juice for brightness and a little sea salt and plenty of freshly ground black pepper.

Arrange the cima on a warm plate and spoon the agresto on top. Serve immediately. This is lovely as a vegetable dish on its own or served alongside grilled meat or fish.

Grilled mackerel with roasted winter tomatoes & almonds

Look for the freshest mackerel available. The flesh should be firm, glossy and smell sweet. The eyes should be clear and bright. Ask your fishmonger to fillet it for you. Mackerel is a wonderful fish, sustainable and abundant in the UK, whilst also being high in omega oils. Mackerel contains both omega-3 and vitamin D – these help you produce serotonin, the happy hormone.

6 winter tomatoes
olive oil, for drizzling
a slice of coarse peasant-style bread
1 garlic clove, peeled
a handful of almonds, skin on
a bunch of parsley, leaves only
3 sprigs of marjoram, leaves only
1 tablespoon red wine vinegar
4 tablespoons extra-virgin olive oil
4 mackerel
sea salt and pepper

SERVES 4
Cooking time 45 mins

METHOD
Preheat the oven to 160°C (320°F), Gas Mark 3.

Slice the tomatoes or leave whole if they are small. Place in a roasting pan and drizzle with a little olive oil. Season with salt and pepper, place on the middle shelf of the oven and roast for 20 minutes or until the tomatoes have begun to soften and ooze a little of their juice. Remove and set aside.

Set your grill (broiler) to high. Toast the bread on both sides and when golden, remove from the grill. Rub both sides with the whole garlic clove, drizzle with olive oil, season with a little salt and tear into small pieces. Place in a bowl along with the tomatoes and their juice.

Roughly chop the almonds and add to the bowl along with the parsley and marjoram. Add the vinegar and toss together to combine. Finally drizzle over the olive oil and set aside while you grill the fish.

Season the mackerel generously with salt and pepper, skin-side only. Place the mackerel on the grill, skin-side down and cook for 3 minutes without turning. The skin should be deliciously crisp and the flesh just opaque. Remove carefully. Arrange on warm plates and spoon over the sauce.

Sour cherry juice

Both sweet and sour cherries have a very short season – they are around for approx. a month in early summer. Like all fruit, when they finally do arrive it tends to be in large quantities rather than a steady trickle. To counteract the gluts, I like to freeze some fruit to capture their flavour and goodness as soon as they are picked, to be enjoyed later in the year when fruit is scarce. Tart cherries are very high in the hormone melatonin and, believe it or not, there have been studies on adults who consume their rich, sharp juice that conclude that cherry juice increases sleep time and sleep efficiency.

2kg (4lb 8oz) tart cherries
2 litres (70fl oz) water
340g (12oz) raw honey, plus extra to serve

SERVES 4 (undiluted) or 12 (diluted)
Cooking time 40 mins

METHOD
Wash and clean the cherries. Remove the stems but leave the pips, as they will add a pleasing almond flavour. Combine the cherries with the water in a deep pan and bring to the boil. Turn down the heat slightly and simmer for 20 minutes. Drain through a fine-mesh strainer, pressing down on the fruit to release as much juice as possible.

Pour the strained juice back into the pan and add the raw honey. Turn the heat up and simmer for 10 minutes – it should reduce and thicken slightly.

Remove from the heat and pour into a sterilized bottle. Store in the fridge for up to a week.

To serve, warm gently and serve with a spoonful or so of extra honey for a little added sweetness.

Almond milk panna cotta with honeycomb & chamomile

Here I have used chamomile as it is the flower that I would traditionally relate to sleep for its calming effect. Almonds are a great sleep-inducing nut, as mentioned before, for their winning combination of zinc, magnesium and melatonin. Milk is also very high in melatonin, so if you want to make a milk panna cotta, swap out the almond milk for organic cow's milk. And cows that are milked at night are shown to have more melatonin in the milk!

ALMOND MILK PANNA COTTA
¾ teaspoon agar-agar
2 tablespoons water
500ml (18fl oz) almond milk
40g (1½oz) raw caster (superfine) sugar
1 tablespoon vanilla extract

50ml (2fl oz) honey
2 tablespoons water
a few sprigs of fresh chamomile
12 almonds

SERVES 4
Cooking time 20 mins, plus 4 hours setting

METHOD
To make the panna cotta, begin by mixing the agar-agar with the water.

Place the almond milk in a saucepan and, over a low flame, let it come to a simmer. Add the sugar and vanilla extract and bring to a boil, allowing the sugar to dissolve.

Stir in the agar-agar and boil for a further 2–3 minutes. Remove from the stove and carefully pour the mixture through a fine sieve into a clean bowl. Allow to cool slightly, then pour into prepared dariole moulds or little glasses and set in the fridge. They will need to sit for at least 4 hours in order to firm to a good consistency. They should be set but still slightly wobbly.

While the cream is setting, place the honey in a pan with the water and a little chamomile. Bring to a simmer, stirring once or twice, then remove from the stove. Allow to cool before adding the almonds.

To serve, unmould the set creams by inverting onto a plate. Spoon over the honey, almonds and chamomile and serve.

Lemon verbena & chamomile tea

Lemon verbena has calming effects and some research has shown it improved sleep. Here I have coupled it chamomile. Chamomile can help improve your quality of sleep, as well as reduce insomnia and chronic inability to sleep. This is a lovely tea as both herbs are easy to cultivate at home, pick and dry ready for the infusion.

1½ teaspoon lemon verbena (fresh or dried)
1 teaspoon dried chamomile
2 teaspoon honey
1 litre filtered water

SERVES 4
Cooking time 10 minutes

METHOD
Bring the water to the boil on the stove before adding in the lemon verbena and chamomile.

Reduce the heat and let simmer for a further 5 minutes.

Stir in the honey before filtering out the herbs with a fine sieve. Then transfer to a teapot and serve.

It is often at this time of year that our minds turn to the idea of community. You might have walked past someone sleeping rough or (depending on where you are) Christmas may be upon you. The cold weather focuses our attention on the people that are less fortunate than ourselves. Shelter, a homeless charity, estimated that in the winter of 2019, there were 280,000 homeless individuals in England. Often homelessness is out of sight and therefore out of mind. As a society, we haven't just lost touch with people who in one way or another have 'slipped through the net', we have also become disconnected from our own communities. Cultural systems have pushed us to live individualistic lives. The autonomy that we strive for in our consumer-led reality doesn't allow nourishment of the self. When happiness is driven by monetary gain, we find ourselves feeling unfulfilled and disconnected. In my conversation with Bruce Parry that follows, we discussed his experiences within indigenous communities, where accountability is swapped for the idea of 'freedom' and how this is relevant to the modern individual. If we are only striving for the good of ourselves, our community mindset is no longer relevant and structures such as family start to break down. Our clans have long since evaporated, and the strains of modern society, coupled with our complex lives, no longer accommodate our ageing. There are 1.4 million older people in England that report chronic loneliness. Once the pillars of wisdom within the community, now they are the fuel for the lucrative care home industry. Younger people are also reporting a rise in loneliness and depression. Society is so intricately linked through social media and the web, yet we have become detached from ourselves and our surroundings. Looking out at the world via some Insta-celebrity's lens can make our own lives feel dull and isolating. Food culture has also become an individual experience, from the lonely takeaway

THE RISE OF THE INDIVIDUAL

to the interruption of a meal by a quick culinary social-media update. We are no longer present. Hooked by dopamine and unable to resist our screens, even at the table, we allow ourselves to be distracted by our pocket companion. These are vital moments when one can be checking in with kin, but if the concentration is always broken, can you really connect? As humans, we developed our cognitive function by being social animals. Compared to other mammals, we were not the fastest or the strongest, so we had to use our social skills to fight our way to the top of the food chain. Breaking bread is a remaining link to our communal past. Eating three times a day creates twenty-one opportunities a week to prepare nourishment for yourself, your family, your friends and your local food bank (if you so

wish). Acts of kindness, such as creating a meal for someone, release the hormone oxytocin for you, the person involved and even anyone witnessing it. The act of generosity is infectious – the kinder you are, the more oxytocin you release and, in turn, the kinder you are. Whilst on cloud nine from all your good deeds, oxytocin also boosts your immune system, creativity and your problem-solving abilities. It also reduces your need for a hit of that addictive hormone – dopamine. Win, win. No matter how big or small your community is, reconnecting is key, be it through sharing food and conversation or by contributing to community projects.

A conversation with

BRUCE PARRY

We were in the first lockdown, back in March 2020, when I talked to explorer Bruce Parry. If there is any man equipped for survival in a global pandemic, it is him, and I would very much like to be in his tribe. Bruce is best known for his expeditions to indigenous communities around the world in the form of his BBC documentary *Tribe*. A former Royal Marines Commando Officer, he now advocates for indigenous people and their land through activism, documentaries and divulgation. He has made a study on community and co-authored a manifesto called 'Anima's Way', hoping to realise it in the form of a commune in Wales. As he's hiding out in the Welsh hills, we connect via FaceTime. This is one of many conversations that I've had with Bruce and from the moment I met him, his inquisitiveness for life and thought has drawn out of me my own ideas and interests.

AB   Anna Boglione
BP   Bruce Parry

AB I want to discuss the sentence or phrase that you use in your manifesto 'Anima's Way', 'the cult of the individual'. We can loop back to 'Anima's Way' after. For now, I really want to talk about the rise of individualism and what you see that to be, and how you see that to have affected our engagement with the idea of community and family? Whether family can recreate a community environment? Has capitalism in the western world pushed the rise in individualism and the want to be free?

BP Um, wow, not a little question. It's quite hard living in a community. When I go and visit an indigenous group in the middle of nowhere, there's a lot of harmony and often a lot of joy and beauty in those groups. When a tribe brushes up against the modern world, modern as in the industrial, technological world that we live in, it's very often the case that the young people, especially the young men, will rush towards our world because it seems to offer so much. One of those elements is the idea of freedom. The idea that you don't need to be constrained by your past, the elders or the norms of the group; you can actually be and do whatever you want. The idea that money can give you the freedom to be all these various things, this allows the individual to be untethered from the confines of living in the community, allowing you to slide into this realm of being an autonomous individual and having all these freedoms. What you notice is that with this movement towards more choice, towards more perception of freedom, there actually comes a huge cost. You are no longer held accountable. You're free to be whatever you want to be. What happens when you start on that path is this sort of movement into separation from your community; a disconnection from the unity of a group with a singular value system. Then you enter into the realm of being able to be 'free' and, of course, behaviours change. On that journey, ownership and hierarchy come into action because people are living more individualistic lives.

As those bonds were broken in the past, people embarked on a journey that highlighted the idea of freedom and individualism as the perfect dream, goal or destination for us all. So much so that even the Constitution of America, the Declaration of Independence, puts the pursuit of freedom and happiness as the highest inalienable rights that humans can have. Which in its own right is almost a statement because it's an individual freedom. It's a statement of individualism, that very thing. The cult of the American dream has also swept the globe. It's a false idol in my view because what we notice is that the more I am on my own, the less I'm held accountable and the more I don't have to deal with my turmoil, so to speak.

That's why I mentioned the cult of the individual. We want it and, as a species, have become so successful at getting what we want. We don't realize that actually what we want isn't very good for us. To say that the pursuit of freedom is a false idol or false goal is beyond where most people are even thinking. They think, 'How could you say that – freedom is the ultimate idea?' In my experience, tribes who have the means and tools for holding everyone to account within society are actually much happier. It's like a child that you let go free and he just becomes a crazy, mad kid, experiencing tantrums, feeling isolated and frustrated. Boundaries are good for us, and boundaries are held within community.

AB Accountability is key and also the rise in freedom means that we are striving after our own individual happiness, which a lot of the time is driven by monetary gain. We put monetary values on happiness and we are setting ourselves up for failure.

BP The pursuit of individual happiness is almost guaranteed to bring misery. The pursuit of meaning, I mean, we talked about this yesterday when we talked about Viktor Frankl. It's such a valid point he made. If you're pursuing happiness, you're going to have really difficult days, but if you pursue meaning…

AB   Do you think we need to define or redefine the idea of freedom? Obviously, freedom is such a huge word, wars have been forged over freedom.

BP   What I advocate is full freedom. I'm an anarchist, if I'm honest. I believe in it as a political state of being. The anarchists that I've lived with are the most peaceful people on the planet. Wait till you watch my film *Tawai*; you'll hear my friend Jerome say, 'This group of people are the most peaceful people on the planet.' I remember saying to him, 'How can you say that, mate? We're making a film. People are going to watch this. I can't just come out with ridiculous statements like that.' He said, 'No Bruce, I'm the head of anthropology at UCL and people have been studying these types of groups now for decades. Time and time again, they come out on top by a long, long, long margin as the most peaceful people on the planet and there they are, they're anarchists.' We've gone so far the other way that we've almost added negativity to every one of these aspects, which were once our way of being. It's down to power; it's down to the competition for narratives that we've been playing for a long, long time and it's down to beliefs. We've just forgotten our true paths. If you are a fully anarchical individual being, as you could be in one of these tribes, that comes with a responsibility. Anarchy is not just throw the TV out of the hotel room, as we think anarchy is; all anarchy means is no leader. 'Monarchy' is one leader; 'anarchy' is without a leader. That's what it means. So, anarchy is where everyone is a fully autonomous individual within society, with the same voice and the same amount of power as everyone else. No one has to be on tiptoes because of stature or power. Everyone is just themselves, so there's full freedom. When you have that much freedom, there is a realization that actually we're better off if we pull together. You choose to be you, choose to be held accountable, that's the beauty of it. You're not forced, you choose it. That's the difference that people don't get. It's when you're that free, unhinged and unlimited by the pressures of society. This won't come overnight for us because we're so wounded. When you see that you've gone on that journey and you've come out the other side, then you're not caught up in the whole ego battle, the status battle and the who am I? You can be of service to the whole now. This is okay; this is actually much cooler and calmer and easier for me. My inner well-being is directly related to the outer well-being that I find myself in, with the people around me and also the environment that supports us. That means that I'm gonna be of service to the people and the place; that just makes sense for me. That is a reality for those people and could be for us.

AB   How do you think we make it a reality for us, for people who are still living very individual lives? Especially people of my age, how do you unite them on the path of connectivity and consciousness that holds them accountable?

BP   Not easily, is the honest truth. The first thing that has to happen is an awakening to the new narrative because at the moment we're all still being sold the narrative of freedom, goods, possessions, power, looks and all of these things. That's clearly what the power wants us to believe because it actually serves them, so we can be on our gerbil wheels running around fuelling the fire of consumerism. 'If only I work harder, then I'll be happy.' If you ask anyone on their death bed what their regrets are, you'll find that they wish they hadn't spent so much time trying to strive for these various things when, actually, true happiness comes from all of those simple aspects of life: being with family and friends under a tree, by a river. We've been sold the wrong narrative.

It's hard to give up the freedom and to step into this unity way of being. What's asked of us is to think and feel differently, and that generally means starting to unpack the stuff that's inside and allowing things to surface. One of the reasons that we like our freedom is because it means, 'I am not accountable'.

This is why I was talking about children being in the community and having the ability to have somewhere to go next door to offset all the stress that they're receiving from their parents. We as a society have got into more separated family units, but we're actually much more wounded than the generations who were living in more open communities. We've all received the trauma from our parents because there's been no escape valve for us. And we carry that. That's why we spend so much money on therapy and going to the jungle or whatever it is, you know, because we're so wounded.

It's actually a lot easier to stay in that place, stay in denial and stay busy and not have to deal with it. When you start coming back into this communal way of being, the first thing that happens is people start going, 'You're being a dick, you know', and it's much easier to avoid that. The only way that we're going to come back into a communal way of being is by accepting that it's a necessary journey because it will bring greater happiness in the long run, but the steps to get there are going to be painful. We're going to have to look at ourselves and go, yeah, I've been behaving in all sorts of antisocial ways and I had no idea. I thought I was being nice, but actually no one was telling me because I was rich or I was successful in life.

AB　You go home to your own house and no one is going to tell you. If you're just interacting with your friends at the pub, they're not going to want to get into a conflict with you. Maybe it's the first time that they've seen you in a couple of weeks, so they just let it fly. They just brush over it and continue.

BP　Then you start living with someone, and then who's doing the washing up? Are you pulling your weight in the garden, or whatever it is? When you're in a truly egalitarian space and there is no power, or everyone's working to keep the power level, then you suddenly realize, okay, this is a totally different space.

That's a journey that not a lot of people actually want to do. I mean, I saw that when I was suddenly on television and I went from Mr happy-go-lucky, glass-half-full, to everyone wanting a bit of my attention. You know, you start changing and there's power and you can use it for manipulation. If someone is making you feel a bit uncomfortable, you just choose not to be with them because you don't need to be. All these things are what will come up when you start moving into connection again.

AB　Also, power feeds the ego, so we're talking about maybe feeding the self rather than the ego. The ego feels great. Living on your ego, it's like living on a cloud of beautiful hot air, you're just floating through life, and the self is actually the person that's on the ground, with their feet in the earth, going, 'Excuse me, can you come back?' But the self actually feels better, the self is substance.

BP　Totally, and you know, we are nourished better with really wholesome interactions, rather than these more shallow, frivolous, playful and momentarily joyful ones. It's a hollow form of happiness, which is almost addictive – you need more and more and more. Rather than sinking into something richer and deeper, where you can slow down in life and not consume as much as well. One of the reasons we're eating the planet is because we're missing something inside. The thing that we're missing is the connection. We're missing connection toward our deeper selves and we're missing connections to nature and to community and to others.

PETERSHAM NURSERIES

Growing up in and out of the nurseries was at times like having our own secret garden, learning what tastes, smells, colours and textures go together. Although at other times it was my delinquent punishment. When we first opened with Skye at the helm, I had dropped out of school and spent my days as their very first kitchen porter. Nowadays, I walk into the greenhouses appreciating how my senses are enraptured by each season and everything it has to offer. The subtleties of my mother are found in the details, from the produce to the decor and within the warmth of the team. She has somehow extended her Aussie heartiness to create an extended family comprised of house guests, the Petersham team and customers. For this reason, I paired Petersham Nurseries with this chapter. From the workshops to the outreach programs, there is always a hustle of people creating together and striving toward common goals.

The chefs create an exciting ever-changing menu, sticking tightly to the ethos of Petersham Nurseries: slow food, great produce, home-grown, healthy, delicious and beautiful. The dishes that come out are vibrant and colourful, incorporating edible flowers from the kitchen garden to match the foliage in the restaurant.

Steamed clams with butter beans & coastal vegetables

These days you want to think before you buy seafood. Where does it come from, how has it been harvested or caught? With clams, you can put your mind at ease – they are both sustainable and nutritious. Clams are high in omega-3s, whilst having very low mercury levels.

Mercury is something to be aware of in larger fish. The higher up the food chain, the more mercury the fish will have. Clams add to their community, by filtering up to 50 gallons of water a day, creating a cleaner habitat for the other fish. We can take a little inspiration from these shelled creatures, and volunteer at our own community clean-ups!

600g (1lb 5oz) butter beans
1.2kg (2lb 10oz) clams
100ml (3½fl oz) olive oil, plus extra
 for drizzling
20g (¾oz) chilli, diced
2 garlic cloves, crushed
175ml (6fl oz) white wine
500ml (18fl oz) stock (fish or vegetable)
180g (6½oz) sea vegetables (agretti,
 samphire, sea beets)
30g (1oz) butter
juice of ½ lemon
50g (1¾oz) garden herbs
 (mint, parsley, chives, oregano),
 chopped

6 slices of focaccia (see page 257) or
 gluten-free seed bread (see page 258)

SERVES 6
Cooking time 1 hour

METHOD
Pre-soak the butter beans the night before in cold water. Simmer them in seasoned water from cold, for around half an hour until cooked. Leave in the water until added to the clams.

Rinse the clams in cold running water for 5 minutes and discard any open or broken shells.

Pour the olive oil into a large pan, heat up and sweat the chilli and garlic. Add the clams and white wine, then cover with a lid for 1 minute.

Add the stock, sea vegetables and cooked butter beans and simmer for 2 minutes. Finish with butter and the lemon juice.

Serve with toasted or grilled slices of focaccia and top with the mixed chopped herbs and a drizzle of olive oil.

Gluten-free seed bread

This is an easy, delicious alternative to bread for those who want to avoid gluten, packed full of omegas, vitamins and antioxidants. This bread is full of fibre and great for your digestive system!

135g (5oz) sunflower seeds
90g (3¼oz) flax seeds
65g (2¼oz) hazelnuts or almonds
145g (5oz) rolled (porridge) oats
   or buckwheat flakes
2 tablespoons chia seeds
4 tablespoons psyllium husk powder
2 teaspoons sea salt
2 tablespoons maple syrup or honey
3 tablespoons any oil or softened butter
350ml (12fl oz) water

MAKES 18–20 SLICES
Cooking time 50 minutes

METHOD
Mix all the ingredients very well in a bowl until everything is combined. If the dough is not combining it may need a little more water, add it in slowly until the dough keeps a round shape without any cracks.

Place in an oiled and lined loaf or cake tin then smooth out the top with the back of a spoon. Let sit for 2 hours or overnight.

Preheat the oven to 175°C (350°F), Gas Mark 4.

Bake for 20 minutes, then remove from the tin, turn over and cook for another 30 minutes. Try and leave to cool completely.

This lasts a week and is good for toast. You can also cut it into slices and freeze.

Focaccia

Focaccia is a really fun and easy bread to make. Being versatile and light you can try adding different toppings to the bread. Rosemary and thinly sliced shallots is always delicious.

222g strong white flour
5g sea salt, plus more to sprinkle
5g fresh yeast
25ml (1fl oz) olive oil
142g room-temperature water

MAKES 8–10 SLICES
Cooking time 30 minutes

METHOD
In a standing mixer bowl, with the hook attachment, add the flour, then place the salt on one side of the bowl, the yeast on the other, oil and water in the middle. Mix well on a medium speed for 6–8 minutes.

Stretch the dough by hand in the bowl, tuck the sides into the centre, then turn the bowl 90 degrees and repeat the stretching, folding and turning process for 5 minutes.

Tip the dough into an oiled bowl, cover and leave to prove for at least one hour.

Preheat the oven to 220 °C (425°F), Gas Mark 7 and line a tray with parchment paper. Add the dough to the tray and give it another hour to prove. Using a finger, dimple the focaccia and sprinkle with sea salt. Put it in the oven.

Turn the oven down after 10 minutes to 200°C (400°F), Gas Mark 6 and cook for another 20 until golden.

Seared mackerel with a quick pickle & crispy capers

We've chosen mackerel because as a nation it's one of the most prolific and sustainable fish we have. It also happens to be a superfood because it's very high in vitamin B12 & omega-3 fatty acids, nutrients we seem to struggle to consume naturally these days, especially in the winter months.

We've paired this stunning fish with lightly pickled vegetables, which will cut through the oiliness of the mackerel, and some Italian radicchio. All the ingredients will be in season through the winter months, bringing brightness in colour and flavour to your meal and giving your body a good dose of some essential vitamins and omega 3.

If you ask your fishmonger to butterfly cut your fish for you, which is a tricky task if you haven't faced it before, he will remove the bones, leaving the tail attached. This is a great way to cook it as it will keep the fish intact in the process and it looks a lot more striking on the plate than fillets would.

100g (3½oz) capers
rapeseed oil
150g (5½oz) shallots
200g (7oz) red/orange carrots
200g (7oz) yellow carrots
220g (7½oz) watermelon 'red meat' radishes
6 mackerel, butterflied
250g (9oz) radicchio Grumolo
250g (9oz) Castelfranco radicchio
250g (9oz) Rosella di Lusia radicchio
extra-virgin olive oil
picked fresh herbs, to garnish
salt and pepper

PICKLING LIQUOR
350ml (12fl oz) red wine vinegar
350ml (12fl oz) white wine vinegar
600ml (20fl oz) water
1 sprig thyme
1 pinch dried chilli
30g (1oz) brown sugar
peel of 1 orange
1 teaspoon fennel seeds
2 star anise

SERVES 6
Cooking time 30 minutes

METHOD
To fry the capers, first make sure you have drained them and dried them on a towel. Heat up a small pan of rapeseed oil to 180°C (356°F) and drop the capers into the oil. They should sizzle and pop open like little flower buds. Remove from the oil with a slotted spoon and dry on a tray with a cloth.

Peel the shallots and carrots, wash the radishes, then slice the shallots into thin rounds, the carrots on an angle and the radishes into half-moons, all as thinly as you can, with a knife.

For your pickling liquor combine all the ingredients and bring to the boil to infuse. Pass the liquor through a fine-mesh sieve and pour over the shallots, carrots and radishes – it should just cover the top. The liquor needs to be near boiling. Once cool, strain, remove the vegetables from the pickling liquor and place the veg into a mixing bowl.

Heat a non-stick pan and season the mackerel with salt, pepper and rapeseed oil. Oil the hot pan and place the mackerel in away from yourself, skin-side down, and press the flesh of the fish

down for the first 10 seconds to stop it curling up. Cooking time will depend on the size of the fish. For a medium butterflied mackerel, it should take around 3 minutes, skin-side down, 2 minutes on the other side. Remove from the heat. It will finish cooking in the pan in the residual heat whilst you finish the salad.

To finish the salad, add the mixed pickled veg into a bowl and mix with the Grumolo, Castelfranco and Rosella di Lusia radicchio (or you can use any seasonal radicchio/endive that you can source), olive oil, salt and pepper and picked fresh herbs (we used parsley).

Plate the salad with the mackerel and finish with a sprinkle of crispy capers.

Roast guinea fowl with butternut squash, cavolo nero, pancetta & crispy rye with a walnut salsa verde

1 guinea fowl or chicken crown
olive oil
½ lemon, plus extra juice for the cavolo nero
10g (¼oz) rosemary sprigs
150g (5½oz) cavolo nero
sea salt and pepper

WALNUT SALSA VERDE
20g (¾oz) oregano
30g (1oz) parsley
30g (1oz) three-cornered leeks
100g (3½oz) young wet walnuts in their shell
2 teaspoon lemon juice
50ml (2fl oz) olive oil
1 teaspoon salt

BUTTERNUT SQUASH PURÉE
1kg (2lb 4oz) butternut squash
300ml (10fl oz) chicken stock
75g (2½oz) butter

CRISPY RYE CROUTONS
200g (7oz) rye bread
200g (7oz) pancetta
30g (1oz) butter
pinch of sea salt

I've chosen this recipe, which we've got on the menu in the Café at Petersham Nurseries, because it's such a winter warming dish, which is really wholesome, super-seasonal and can be enjoyed from one person all the way to a large party/banquet. The colours and flavours of the dish are very vibrant and it will definitely be a crowd-pleaser.

The elements of the dish can all be adapted depending on what you can get hold of. We've used guinea fowl for this, but chicken or partridge would work equally well. Likewise with the vegetables, where you can use whatever squash or pumpkin you can buy locally.

Three-cornered leeks are part of the allium family, which grow wild and abundantly from November all the way through till January and can be easily identified by their three corners in the grass-like leaves (hence the name) and garlicky smell. A good walk in woodland or a large park will likely find them where they have been left to grow wild – just be sure to give them a good wash before use.

SERVES 2–4
Cooking time 1 hour 45 minutes

METHOD
Start by preheating your oven to 200°C (400°F), Gas Mark 6.

Give the guinea fowl a coating of olive oil and season with sea salt and cracked black pepper. Stuff the cavity with the lemon and some sprigs of rosemary and roast for 40 minutes, allowing 30 minutes to rest after cooking.

For the walnut salsa verde, pick the leaves from the oregano and parsley and wash the three-cornered leeks, then finely chop all of the herbs.

Shell the walnuts and roughly chop them, then combine in a bowl with the chopped herbs, lemon juice, olive oil and salt. You want to serve this at room temperature, so make sure it's out of the fridge for an hour before serving.

For the butternut squash purée, peel and deseed the squash (you can keep all the trimmings to make a stock or a broth). Chop up the squash into 4cm (1½ inch) pieces, roughly all the same size so they cook evenly, and simmer the squash in the chicken stock and butter, seasoned with

salt and pepper, for about 15 minutes, until it's soft and you can poke a skewer or a knife through it easily. Let it cool down for about 15 minutes, then blitz until as smooth as possible in a blender.

For the rye croutons, dice up the rye bread into small 1cm (½ inch) cubes. Do this by cutting even slices and then stack them up like a pack of cards and slice 1cm (½ inch) soldiers. From there, turn them horizontally and cut your even cubes. You don't want to try to do too much at the same time, so start with three or four slices and see how you go.

Cut your pancetta into slightly larger cubes and start frying in a pan. When they're starting to crisp and sizzle away, add your butter and rye bread and keep the pan moving regularly until the rye bread is crispy. Drain all of them onto a cloth ready to serve and keep in a warm place or warm up before serving.

Pick the cavolo nero from the stalk and boil in salted water for 3–4 minutes. Drain well, tip straight into a hot pan with olive oil and add a squeeze of lemon and a pinch of salt before serving.

Nettle risotto & toasted hazelnuts

Nettles are good for an array of different things. Being high in nitrogen and good for composting, they can be turned into plant food. Stinging nettles are also great for human consumption. They are packed with nutrients, vitamins C and A, calcium, magnesium, iron and potassium. They are also a great source of protein and have pain-relieving and anti-inflammatory properties, known to help arthritis and relieve inflammation. Nettle tea is commonly used to reduce joint inflammation.

NETTLE PURÉE
80g (2¾oz) nettles
4 teaspoons vegetable stock
large pinch of sea salt

RISOTTO
1.2 litres (2 pints) vegetable stock
2 tablespoons olive oil
150g (5½oz) shallots, diced
200g (7oz) risotto rice
100ml (3½fl oz) white wine
50g (1¾oz) butter
90g (3¼oz) Parmesan, grated
100g (3½oz) hazelnuts, toasted
  and chopped
salt

MAKES 6 PORTIONS, AS A STARTER
Cooking time 35 minutes

METHOD
Blanch the nettles in salted boiling water for 30 seconds, then plunge into ice water and blitz into a purée with the 4 teaspoons of vegetable stock and salt.

Heat the stock in a small pot. In another medium pot, heat the olive oil and sweat the shallots until soft. Add the rice, combine the ingredients and season with salt.

Add your white wine and stir until the liquid has disappeared, then add the stock, a ladle at a time, whilst constantly stirring the risotto until it's soft with a slight bite to it.

At this stage, add your nettle purée and combine. Take off the heat, add the butter and Parmesan to the risotto and check the seasoning. The risotto is ready to serve, finished with chopped hazelnuts over the top.

PLANT FOOD
Fill a bucket with nettles, top the bucket up with water and leave for 10 days. Dilute with water – 10 parts water to 1 part nettle juice – there you go, a cost-effective organic fertilizer.

NETTLE TEA
Add a handful of leaves to water and bring the water to a boil. Turn off the heat and let the tea sit for 5 minutes. Add honey, lemon or cinnamon, to taste. You can also experiment by using other plants, such as mint and chamomile.

Vegan Medjool date & orange cake

We had to include a cake! Bruce talks about egalitarian communities sharing everything equally. We might not be able to include their ethos into all of our day-to-day lives, but where we can, we should. A gluten-free, vegan cake is a good place to start. Cut it into equal slices and share with your community, new and old.

150g (5½oz) Medjool dates, chopped, just covered in boiling water and left to soak
350g (12oz) plain (all-purpose) flour or gluten-free flour
225g (8oz) dark brown sugar
100ml (3½fl oz) sunflower oil
zest of 1 large orange
1 teaspoon baking powder
½ teaspoon bicarbonate of soda (baking soda)
pinch table salt
4 tablespoons maple syrup

CANDIED ORANGE SLICES
oranges
water
sugar

FOR THE ICING
225g (8oz) icing (confectioner's) sugar
25g (1oz) almond butter
1 teaspoon vanilla extract
2 tablespoon almond milk

SERVES 8
Cooking time 30 minutes

METHOD
For the candied orange slices, you can make as much or as little as you like. They will keep for a month in an airtight container and are great to have around Christmas time. To make, slice your oranges ½cm (¼ inch) thick. In a pan, combine equal weights of water and sugar – use as much as you need to cover the orange.

Bring the syrup to the boil and then turn the heat down to a gentle simmer. Add the orange slices and simmer, stirring every 15 minutes. Drain and leave to cool on a wire rack – all separate and not overlapping – for up to 24 hours and then dip into granulated sugar or melted chocolate. Don't discard the cooking syrup; you can use it again or to make jams/marmalades.

Drain the dates and keep the soaking water – you'll need 200ml (7fl oz) in total of water, so top it up if you need to.

Blend 50g (1¾oz) of the dates until smooth and add to a bowl with the remaining dates and all the other ingredients. Mix until you have a smooth batter.

Preheat the oven to180°C (356°F). Pour into a 20–25cm (8–10 inch) cake tin lined with parchment paper and bake for 25–30 minutes until fully risen – when you poke it with a skewers, the skewers should come out clean.

To make the icing, add the sugar, almond butter, vanilla extract and half of the almond milk to a mixer. Starting on a low speed and gradually increasing, add in the rest of the almond milk slowly until perfectly smooth. This should result in a very thick but spreadable icing.

Let the cake cool then ice the top with the buttercream. Finally decorate with the candied orange slices.

INDEX

PICTURE INDEX

THANKS

This has been a real journey from start to finish and many things have been learned along the way. The first Covid lockdown allowed me the time to write the proposal for Recipes To Reconnect. My close friend and life-ally, Rufus Besterman, had moved in to weather the storm. He diligently read every word, cleaned up my thoughts and challenged my ideas. Later this role would fall to Riccardo Basile, my love, who took to it with ruthless passion. Winter lockdown was spent sharing ideas and laughing at my interview hiccups, he taught me more about my self and my writing than anyone else to enter my life. Without him, I'm not sure if this book would have been possible. Beyond these two very special people in my life, I would like to thank my family and friends who kindly read sections when I was struggling, my little brother Harry for always sharing his insight and intelligence with me, and my besties Clementine King, for reading and critiquing, and Billie De Melo Wood, for elegantly breaking down complex ancient yogic teachings to me. There are only a few people I felt comfortable enough to share my work with, so thank you for your time and honesty.

I have deeply enjoyed my working relationship with Joanna Copestick and Isabel Jessop at Kyle Books, we have had synergy throughout, I can't thank them enough for the support. To Joanna for cleverly suggesting how to improve the book and to Isabel for always being on hand to help figure things out, thank you both. Thank you also for introducing me to the graphic designer Carol Montpart, who took a deep interest in the book and made it everything I'd hoped for. Carol, it has been an absolute pleasure to work with you and your team. To my literary agent, Caroline Michel, thank you for taking me on and believing in the project. I look forward to our relationship blossoming into the future.

Finally but most importantly, thank you to each person that lent their time, words and recipes: without you believing in this project it would have not been possible. Seeing and hearing about the wonderful work each of you is doing has opened my heart, and restored my faith in humanity. With minds and people like you we can help to heal this planet and in turn, the people that reside here.

My deepest thanks